Introduction to Cable Television (CATV):
Analog and Digital Cable Television and Modems

Lawrence Harte

2nd Edition

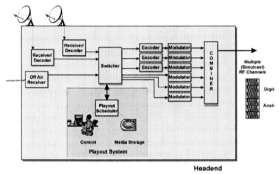

Head End System

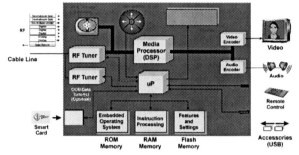

Set Top Box Operation

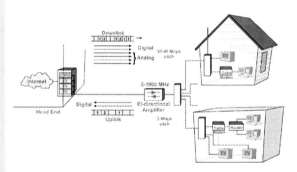

Cable Modem System

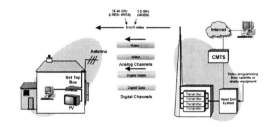

LMDS Radio

Excerpted From:

CATV Systems

With Updated Information

ALTHOS

ALTHOS Publishing

ALTHOS Publishing

Copyright © 2007 by the ALTHOS Publishing Inc. All rights reserved. Produced in the United States of America. Except as permitted under the United States Copyright Act of 1976, no part of this publication may be reproduced or distributed in any form or by any means, or stored in a database or retrieval system, without prior written permission of the publisher.

ISBN:0-9728053-6-2

All trademarks are trademarks of their respective owners. We use names to assist in the explanation or description of information to the benefit of the trademark owner and ALTHOS publishing does not have intentions for the infringement of any trademark.

ALTHOS electronic books (ebooks) and images are available for use in educational, promotional materials, training programs, and other uses. For more information about using ALTHOS ebooks and images, please contact us at info @Althos.com or (919) 557-2260

Terms of Use

This is a copyrighted work and ALTHOS Publishing Inc. (ALTHOS) and its licensors reserve all rights in and to the work. This work may be sued for your own noncommercial and personal use; any other use of the work is strictly prohibited. Use of this work is subject to the Copyright Act of 1976, and in addition, this work is subject to these additional terms, except as permitted under the and the right to store and retrieve one copy of the work, you may not disassemble, decompile, copy or reproduce, reverse engineer, alter or modify, develop derivative works based upon these contents, transfer, distribute, publish, sell, or sublicense this work or any part of it without ALTHOS prior consent. Your right to use the work may be terminated if you fail to comply with these terms.

ALTHOS AND ITS LICENSORS MAKE NO WARRANTIES OR GUARANTEES OF THE ACCURACY, SUFFICIENCY OR COMPLETENESS OF THIS WORK NOR THE RESULTS THAT MAY BE OBTAINED FROM THE USE OF MATERIALS CONTAINED WITHIN THE WORK. ALTHOS DISCLAIMS ANY WARRANTY, EXPRESS OR IMPLIED, INCLUDING BUT NOT LIMITED TO IMPLIED WARRANTIES OF MERCHANTABILITY OR FITNESS FOR A PARTICULAR PURPOSE.

ALTHOS and its licensors does warrant and guarantee that the information contained within shall be usable by the purchaser of this material and the limitation of liability shall be limited to the replacement of the media or refund of the purchase price of the work.

ALTHOS and its licensors shall not be liable to you or anyone else for any inaccuracy, error or omission, regardless of cause, in the work or for any damages resulting there from. ALTHOS and/or its licensors shall not be liable for any damages including incidental, indirect, punitive, special, consequential or similar types of damages that may result from the attempted use or operation of the work.

About the Authors

Mr. Harte is the president of Althos, an expert information provider whom researches, trains, and publishes on technology and business industries. He has over 29 years of technology analysis, development, implementation, and business management experience. Mr. Harte has worked for leading companies including Ericsson/General Electric, Audiovox/Toshiba and Westinghouse and has consulted for hundreds of other companies. Mr. Harte continually researches, analyzes, and tests new communication technologies, applications, and services. He has authored over 80 books on telecommunications technologies and business systems covering topics such as mobile telephone systems, data communications, voice over data networks, broadband, prepaid services, billing systems, sales, and Internet marketing. Mr. Harte holds many degrees and certificates including an Executive MBA from Wake Forest University (1995) and a BSET from the University of the State of New York, (1990).

Table of Contents

OVERVIEW 2
    BROADCAST TELEVISION 4
    CABLE TELEVISION SYSTEMS (CATV) 5

CONTRIBUTION NETWORK 9
    CONNECTION TYPES 10
    PROGRAM TRANSFER SCHEDULING 12
    CONTENT FEEDS 13

HEADEND 17
    INTEGRATED RECEIVER DECODER (IRD) 18
    OFF AIR RECEIVERS 20
    ENCODERS 20
    TRANSCODERS 20
    RATE SHAPER 20
    CHANNEL MODULATORS 22
    CHANNEL PROCESSORS 22
    CHANNEL SIGNAL COMBINERS 23

ASSET MANAGEMENT 24
    CONTENT ACQUISITION 25
    METADATA MANAGEMENT 26
    PLAYOUT SCHEDULING 26
    ASSET STORAGE 28
    CONTENT PROCESSING 29
    AD INSERTION 29
    DISTRIBUTION CONTROL 31

Copyright ©, 2007, ALTHOS, Inc

DISTRIBUTION NETWORK 32
CORE NETWORK 32
ACCESS NETWORK 34
PREMISES DISTRIBUTION 36

CATV END USER DEVICES 39
TUNERS 41
DISPLAY CAPABILITY 41
SECURITY 41
MEDIA PROCESSING 42
SOFTWARE APPLICATIONS 43
ACCESSORIES 43
MIDDLEWARE COMPATIBILITY 44
UPGRADABILITY 44
MEDIA PORTABILITY 44

CATV ACCESS DEVICES 44
CABLE READY TELEVISIONS 45
SET-TOP BOX (CABLE CONVERTERS) 45
CABLE MODEMS 47
CABLE TELEPHONE ADAPTERS 48

MARKET GROWTH 48
CABLE TELEVISION MARKET 49
CABLE MODEM MARKET 50
CABLE TELEPHONE MARKET 51

TECHNOLOGIES 52
ANALOG VIDEO 52
DIGITAL VIDEO 53
RIGHTS MANAGEMENT 56
CABLE MODEMS 57
HIGH DEFINITION TELEVISION (HDTV) 59
CABLE TELEPHONY 61
WIRELESS CABLE 62

  INTERACTIVE TELEVISION 64
  INTERNET PROTOCOL TELEVISION (IPTV) 66
SYSTEMS **66**
  NATIONAL TELEVISION STANDARDS COMMITTEE (NTSC) 66
  PHASE ALTERNATING LINE (PAL) 67
  SEQUENTIAL COULEUR AVEC MEMOIRE (SECAM) 69
  MOTION PICTURE EXPERTS GROUP (MPEG) 69
  DATA OVER CABLE SERVICE INTERFACE SPECIFICATIONS (DOCSIS)
                               70
  PACKETCABLE™ 71
  OPENCABLE™ 72
SERVICES **72**
  TELEVISION PROGRAMMING 72
  PAY PER VIEW (PPV) 74
  ADVERTISING 75
  HIGH SPEED DATA (CABLE MODEMS) 76
  CATV TELEPHONE SERVICES 78

APPENDIX 1 - CATV ACRONYMS 81

INDEX. 91

Cable Television (CATV)

Cable television (CATV) is a television distribution system that uses a network of cables to deliver multiple video and audio channels. Since 1941, television broadcast services have been providing news and entertainment to listeners without wires. Because television broadcast systems could only provide radio coverage to a limited geographic area (such as a city), people and companies began to be setup systems to deliver television signals in other areas by interconnection cables in 1948. These early analog cable television systems simply captured television signals in one geographic area and retransmitted these television channels in other geographic areas.

Video broadcasting is the process of transmitting video images to a plurality of receivers. The broadcasting medium may be via radio waves, through wired systems (such as CATV), or through packet data systems (such as the Internet). Television involves the transmission and reception of visual images via electrical signals. Video is an electrical signal that carries TV picture information.

For many years, video (television) broadcasters had monopolized the distribution of some forms of information to the general public. This had resulted in strict regulations on the ownership, operation, and types of services broadcast companies could offer. Due to the recent competition of wide area information distribution, governments throughout the world have eased their regulation of the broadcast industry. In 1996, the United States released the Telecommunications Act of 1996 that dramatically deregulated the telecommunications industry. This allowed broadcasters to provide many new services with their existing networks.

In the mid 1990's, a major technology change occurred in the broadcast industry. Television systems began to convert from analog systems to digital systems. The use of digital transmission enabled broadcasters to transmit new types of services along with digital television channel signals. This included data (Internet) and telephone services. The ability to integrate several services into one transmission signal allows the cable television operator to offer these new services without significant investment in new cable systems.

Overview

Cable television systems were created because of the difficulties of high frequency radio communication systems and to allow more television channels to be transferred to consumers than is available in radio frequency channels. One of the first cable television systems occurred in remote valleys in Pennsylvania in 1948. Due to the shadowing (signal blocking) effects of mountains, people living in valleys solved their reception problems by installing antennas top of the hills and running cables to their houses [1]. These early cable systems were simple antennas that used cable connections to retransmit signals to nearby areas.

The use of cable lines to interconnect the antennas in cable TV systems reduced the signal level (signal attenuation). To overcome the signal attenuation losses of interconnection cables, amplifiers are inserted at regular intervals in cable systems restore the original signal levels. Although amplifiers boost the signal levels, each amplifier adds signal distortion (electrical noise). Due to the cumulative effects of amplification noise, the use of amplifiers limits the maximum distance of a cable television distribution system. Each amplifier is installed approximately every 1,000 feet [2].

In the 1950s, cable system operators began experimenting with the insertion of other media sources on television channels that were not available in their local area. This allowed cable systems to offer programming that was unavailable via normal television broadcast. This increased the interest of customers to purchase cable television services.

Each analog television channel uses 6 to 8 megahertz (MHz) of the radio spectrum to transfer both video signal (a majority of the bandwidth) and audio. In the United States, the FCC allocated a frequency range within the **very high frequency** (VHF) radio spectrum (below 300 MHz) to allow up to 12 television channels. To provide additional television channels, the FCC allocated additional frequencies in the **ultrahigh frequency** (UHF) portion of the radio spectrum (above 300 MHz). Channels 14 to 83 were created in the frequency range of 470 MHz to 894 MHz. The frequency bands for television channels 70 through 83 were eventually reallocated (reassigned) for mobile telephone services in the 1980s.

The first "pay-per-view" (PPV) channel was offered by a cable television system in Wilkes-Barre Pennsylvania in 1972 [3]. This was a regional service called home box office (HBO). In 1975, HBO began transmitting nationwide using satellite transmission. Early satellite systems could broadcast up to 24 channels for each satellite transponder. These early systems required the use of relatively large dish antennas 10 meters in diameter and channel required a separate antenna. This limited the initial distribution of HBO to cable networks.

In 1976, cable systems began to use fiber optic cables to carry television signals from the head-end to the neighborhood [4]. Some of the advantages of fiber-optic cable include lower signal losses than coaxial cable, higher data transmission rate and digital signal regeneration. This reduced the number of amplifiers between the head-end and customer from 30 to 40 down to approximately 1 or 2 [5]. Fiber optic cable has much more bandwidth transmission potential than coaxial cable. A typical coaxial cable system can transfer approximately 10 billion bits per second (Gbps). A single strand of fiber optic cable (and there are usually many fibers per cable) can carry in excess of 1 trillion bits of data (Tbps) [6]. When transferring information in digital form, the signal is regenerated at each amplifier rather than amplified.

Using MPEG digital video compression, CATV systems can transmit up to 10 video channels in the 6 MHz bandwidth of a single analog television channel [7]. When using the large available bandwidth of coax cable, this can provide more than 1,000 digital video channels to consumers.

Broadcast Television

The technology that is used for television broadcast was developed in the 1940s. The success of the television marketplace is due to standardized, reliable, and relatively inexpensive television receivers and a large selection of media sources. The first television transmission standards used analog radio transmission to provide black and white video service. These initial television technologies have evolved to allow for both black and white and color television signals, along with advanced services such as stereo audio and closed caption text. The television system evolved in a way that maintained backward compatibility with existing television equipment allowing the same radio channels to be used for black and white television and color television services.

Television systems initially used analog television technology. While analog television transmission can provide good video and audio signals, it does not easily allow the sending and receiving of digital data. Several new television broadcasts exist that can deliver high quality video and audio as well as information services using digital signal transmission.

New technologies allow transmission of high-definition television (HDTV). HDTV is the term used to describe a high-resolution video and high quality audio signal as compared to standard NTSC or PAL video transmission. HDTV signals can be in analog or digital form. Digital HDTV systems have the added benefits of providing data and other multimedia services.

Figure 1.1 shows a television broadcast system. This television system consists of a television production studio, a high-power transmitter, a communications link between the studio and the transmitter, and network feeds for programming. The production studio controls and mixes the sources of information including videotapes, video studio, computer created images (such as captions), and other video sources. A high-power transmitter broadcasts a single television channel. The television studio is connected to the transmitter by a high bandwidth communications link that can pass video and control signals. This communications link may be a wired (coax) line or a microwave link. Many television stations receive their video source from a television network. This allows a single video source to be relayed to many television transmitters.

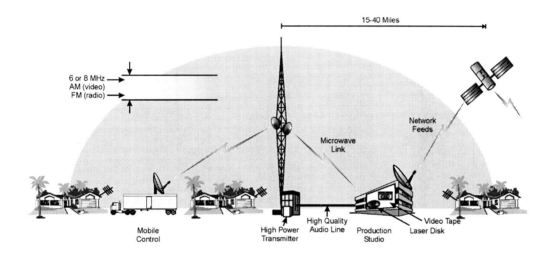

Figure 1.1, Broadcast Television System

Cable Television Systems (CATV)

Cable television is a distribution system that uses a network of cables to deliver multiple video and audio channels. CATV systems can have up to 120 transmission channels. For analog cable systems, each transmission channel provides one video program. For digital cable systems, each transmission channel can provide 4 to 10 television channels. In the late 1990's,

Introduction to Cable Television (CATV), 2nd Edition

many cable systems started converting to digital transmission using fiber optic cable and digital signal compression.

CATV system operators link content providers to content consumers. To do this, CATV systems gather content via a content network and convert the content to a format that it can use via a headend system. It then manages (e.g. playout) the content via an asset management system, transfers the content via a distribution network, and then the media may be displayed on a variety of television devices.

Figure 1.2 shows a sample CATV system. This diagram shows the CATV system gathers content from a variety of sources including network feeds, stored media, communication links and live studio sources. The headend converts the media sources into a form that can be managed and distributed. The asset management system stores, moves and sends out (playout) the media at scheduled times. The distribution system simultaneously transfers multiple channels to users who are connected to the CATV system. Users view CATV programming on televisions that are directly connected to the cable line (cable ready TVs) or through an adapter box (set top box).

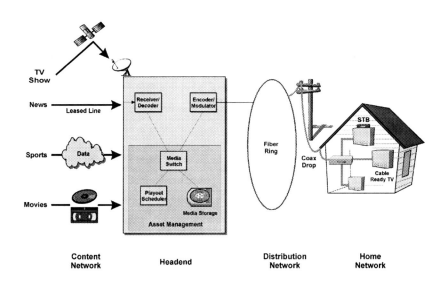

Figure 1.2, CATV System

Cable television systems can be one-way (only from the head-end to consumers) or two-way (both to and from the customer) systems. Initial CATV systems were one way systems that captured broadcasted television programs ("off-air") and simply transferred these signals by cable for retransmission in another geographic area.

Figure 1.3 shows a one-way cable television system. This diagram shows that various video sources are selected in the head-end. Each of the video sources that will be distributed on the cable network is applied to an RF modulator that converts the video signals into RF signals on a specific frequency. The many RF signals are combined (added together), amplified and sent to the cable television system distribution network. The distribution network supplies part of the signal (signal tap) as the cable passes near each home or business location. As the distribution system progresses away from the head-end, the signal level begins decrease. Periodically, amplifiers are used to increase the composite video signal.

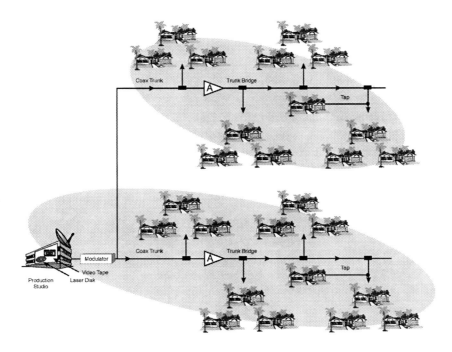

Figure 1.3, One-Way Cable Television System

To provide advanced services, cable systems have been evolving to have two-way capabilities. A two-way cable television system allows customers to receive and send information between the cable system and their set-top box. There are two options for two-way cable television systems: a hybrid system and an integrated system. Hybrid two-way systems use different technologies to transfer information in different directions and integrated systems use the cable network for both directions of communication.

Because the design of most cable systems started as a one-way cable system, hybrid systems were first used to add two-way communication capability. For hybrid systems, a different technology is used to transfer information to the user (downstream) and from the user (upstream). Early systems used the cable for the downstream and the telephone network or wireless data devices for the upstream.

For two-way cable systems that use coaxial (RF) cable, return paths (return transmission channels) can be assigned to frequencies in the range below 50 MHz. This frequency range was unassigned for television operation. For cable systems that have converted their network to use fiber (optical) cable, the systems can use separate fiber strands for each direction as each fiber cable often has several (30+) fiber strands.

The two-way cable system requires cable modems at the user end and a coordinating modem at the head-end of the system. The cable modem is a communication device that modulates and demodulates (MoDem) data signals to and from a cable television system. A modem at the head-end coordinates the customer's modem and interfaces data to other networks (such as the Internet).

Figure 1.4 shows a two-way cable television system. This diagram shows that the two-way cable television system adds a cable modem termination system (CMTS) at the head-end and a cable modem (CM) at the customer's location. The CMTS also provides an interface to other networks such as the Internet.

Introduction to Cable Television (CATV), 2nd Edition

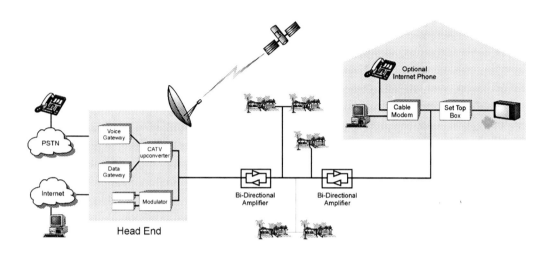

Figure 1.4, Two-Way Cable Television System

A multichannel video program distributor is a company or organization that sells video media services to users.

A multiple system operator is a company that owns more than one telecommunications system that provides communications services. In the United States, MSO is the term that is commonly used to describe a company that owns and operates more than one cable television system.

Contribution Network

A contribution network is a system that interconnects contribution sources (media programs) to a content user (e.g. a television system). CATV systems receive content from multiple sources through connections that range from dedicated high-speed fiber optic connections to the delivery of stored media. Content sources include program networks, content aggregators and a variety of other government, education and public sources.

Connection Types

CATV connection types include satellite connections, leased lines, virtual networks, microwave, mobile data and public data networks (e.g. Internet).

Satellite communication is the use of orbiting satellites to relay communications signals from one station to many others. A satellite communication link includes a communication link that passes through several types of systems. These connections include the transmission electronics and antenna, uplink path, satellite reception and transmission equipment (transponder), downlink path, and reception electronics and antenna. Because satellite systems provide signal coverage to a wide geographic area, the high cost of satellites can be shared by many broadcasting companies.

Satellite content distributors that provide television programming to CATV networks via satellite lease some or all of the transponder capacity of the satellite. Satellite content providers combine multiple programs (channels) for distribution to broadcasters.

Leased lines are telecommunication lines or links that have part or all of their transmission capacity dedicated (reserved) for the exclusive use of a single customer or company. Leased lines often come with a guaranteed level of performance for connections between two points. Leased lines may be used to guarantee the transfer of media at specific times.

Virtual private networks are private communication path(s) that transfer data or information through one or more data network that is dedicated between two or more points. VPN connections allow data to safely and privately pass over public networks (such as the Internet). The data traveling between two points is usually encrypted for privacy. Virtual private networks allow the cost of a public communication system to be shared by multiple companies.

A microwave link uses microwave frequencies (above 1 GHz) for line of sight radio communications (20 to 30 miles) between two directional antennas. Each microwave link transceiver usually offers a standard connection to communication networks such as a T1/E1 or DS3 connection line. This use of microwave links avoids the need to install cables between communication equipment. Microwave links may be licensed (filed and protected by government agencies) or may be unlicensed (through the use of low power within unlicensed regulatory limits). Microwave links are commonly used by CATV systems to connection remote devices or locations such as a mobile news truck or a helicopter feed.

Mobile data is the transmission of digital information through a wireless network where the communication equipment can move or be located over a relatively wide geographic area. The term mobile data is typically applied to the combination of radio transmission devices and computing devices (e.g. computers electronic assemblies) that can transmit data through a mobile communication system (such as a wireless data system or cellular network). In general, the additional of mobility for data communication results in an increased cost for data transmission.

The Internet is a public data network that interconnects private and government computers together. The Internet transfers data from point to point by packets that use Internet protocol (IP). Each transmitted packet in the Internet finds its way through the network switching through nodes (computers). Each node in the Internet forwards received packets to another location (another node) that is closer to its destination. Each node contains routing tables that provide packet forwarding information. The Internet can be effectively used to privately transfer programs through the use of encryption.

In additional to gathering content through communication links, content may be gathered through the use of stored media. Examples of stored media include magnetic tapes (VHS or Beta) and optical disks (CD or DVDs).

When content is delivered through the content network, its' descriptive information (metadata) is also delivered. The metadata information may be embedded within the media file(s) or it may be sent as in separate data files. Some of the descriptive data may include text that is used for closed captioning compliance.

Figure 1.5 shows a contribution network that is used with a CATV system. This example shows that programming that is gathered through a contribution network can come from a variety of sources and that include satellite connections, leased lines, virtual networks, microwave links, mobile data, public data networks (e.g. Internet) and the use of stored media (tapes and DVDs).

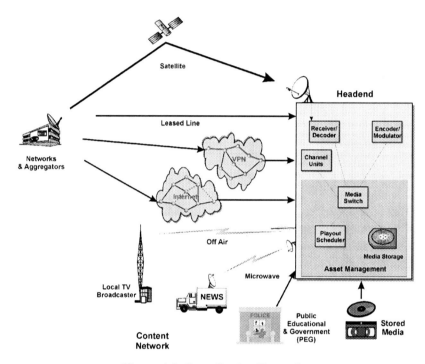

Figure 1.5, Contribution Network

Program Transfer Scheduling

Program transfer scheduling is the setup and management of times and connection types that media will be transported to the CATV system. CATV systems have a limited amount of media storage for television programs so they typically schedule the transfer programming a short time (possibly several days) before it will be broadcasted in their system.

The cost of transferring content can vary based on the connection type (e.g. satellite versus Internet) and the data transfer speed. In general, the faster the data transfer speed, the higher the transfer cost. The scheduling of program transfer during low network capacity usage periods and at lower speed can result in significant reduction in transfer cost.

Figure 1.6 shows how a CATV system may use transfer scheduling to obtain programs reliably and cost effectively. This example shows that the CATV system may select multiple connection types and transfer speeds. This example shows that the selection can depend on the program type (live versus scheduled) and transfer cost.

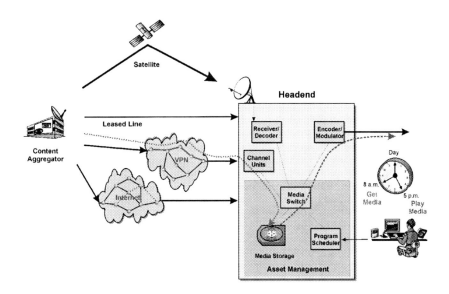

Figure 1.6, CATV Program Transfer Scheduling

Content Feeds

A content feed is a media source that comes from a content provider or stored media system. The types of content feeds that may be used in CATV systems range from network feeds (popular programming) to video feeds from public events (government programming).

Off Air Feed

An off air feed is a content source that comes from an antenna system that captures programming from broadcasted radio signals (off air radio transmission). The off air feed converts broadcasted radio channels into a format that can be retransmitted on another system (such as a CATV system). Off-air feeds are used to retransmit locally broadcasted content on the CATV system.

Network Feed

A network feed is a media connection that is used to transfer media or programs from a network to a distributor of the media or programs.

Local Feed

Local feed is a media connection that is used to transfer content from local sources. Examples of local feeds include connection from sportscasts, news crews and live studio cameras.

Truck Feed

A truck feed is a media connection that is used to transfer content from mobile news vehicle source. Examples of truck feeds include cellular and microwave connections.

Helicopter Feed

Helicopter feed is a media connection that is used to transfer content from airborne sources. Examples of helicopter feeds include microwave and private radio connections.

Live Feed

A live feed is a media connection that is used to transfer media or programs from a device that is capturing in real time (such as a mobile camera) to a distributor of the media or programs.

Government Access Channel

A government access channel is a media source that is dedicated to informing citizens of public related information. Examples of government programming include legal announcements, property zoning, public worker training programs, election coverage, health related disease controls and other public information that is related to citizens.

Educational Access Channel

An educational access channel is a media source that is dedicated to education. Educational programming may come from public or private schools. Examples of educational programming include student programming, school sporting events, distance learning classes, student artistic performances and the viewpoints and teachings of instructors.

Public Access Channel

A public access channel is a media source that is dedicated to allowing the public to create and provide programming to a broadcast system. Examples of public programming include local events and subjects that members of a community have an interest in.

Syndication Feeds

Syndication feeds are media connections or sources that are used to transfer media or programs from a syndicated network to a distributor of the media or programs. An example of a syndicated feed is really simple syndication (RSS) feed. An RSS feed provides content via the Internet such as news stories. RSS allows content from multiple sources to be more easily distributed. RSS content feeds often commonly identified on web sites by an orange rectangular icon.

Emergency Alert System (EAS)

Emergency alert system is a system that coordinates the sending of messages to broadcast networks of cable networks, AM, FM, TV broadcast, low power TV (LPTV) stations and other communications providers during public emergencies. When emergency alert signals are received, the transmission of broadcasting equipment is temporarily shifted to emergency alert messages.

Figure 1.7 shows some of the different types of content sources that may be gathered through a contribution network. This table shows that content sources include off-air local programs, entertainment from national networks, local programs, government access channel (public information), education access, public access (residents), syndication (shared sources), and the emergency alert systems.

Content Sources	Notes
Off-air feed	Local programs
Network feed	Entertainment and nationwide programs
Local feed	Local information
Truck feed	Event and news information
Helicopter feed	Traffic and news
Live feed	Local news
Government access channel	Public information
Education access channel	Schools and learning sources
Public access channel	Community residents
Syndication feeds	News and shared resources
Emergency alert system	Public safety

Figure 1.7, Contribution Network Programming Sources

Headend

The head-end is the master distribution center of a CATV system where incoming television signals from video sources (e.g., DBS satellites, local studios, video players) are received, amplified, and re-modulated onto TV channels for transmission down the CATV system.

The incoming signals for headend systems include satellite receivers, off-air receivers and other types of transmission links. The signals are received (selected) and processed using channel decoders. Headends commonly use integrated receiver devices that combine multiple receiver, decoding and decryption functions into one assembly. After the headend receives, separates and converts incoming signals into new formats, the signals are selected and encoded so they can be retransmitted (or stored) in the CATV network. These signals are modulated, amplified and combined so they can be sent on the CATV distribution system.

Figure 1.8 shows a diagram of a simple head-end system. This diagram shows that the head-end gathers programming sources, decodes, selects and retransmits video programming to the distribution network. The video sources to the headend typically include satellite signals, off air receivers, microwave connections and other video feed signals. The video sources are scrambled to prevent unauthorized viewing before being sent to the cable distribution system. The headend receives, decodes and decrypts these channels. This example shows that the programs that will be broadcasted are supplied to encoders and modulators to produce television channels on multiple frequencies. These channels are combined onto a single transmission line by a channel combiner.

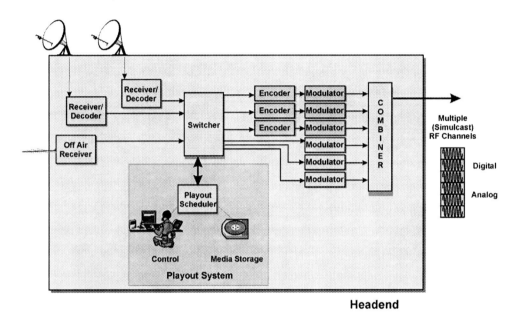

Figure 1.8, Head-end System

A CATV system has expanded to include multiple regions including local headend locations, which may be distributed over a large geographic region. Local headends may be connected to regional headends and regional headends may be connected to a super headend. To reduce the cost of a CATV system, headend systems can be shared by several distribution systems.

Integrated Receiver Decoder (IRD)

An integrated receiver and decoder is a device that can receive, decode, decrypt and convert broadcast signals (such as from a satellite system) into a form that can be transmitted or used by other devices.

In headend systems, IRDs are commonly used to demodulate and decrypt the multi-program transport stream (MPTS) from a satellite antenna. The IRD has a receiver that can select and demodulate a specific channel. The decoder divides an incoming channel into its component parts. A decryptor can convert the encrypted information into a form that can be used by the system. An interface converter may change the format of the media so that it can be used by other devices.

The inputs to an IRD (the front end) can include a satellite receiver or a data connection (such as an ATM or IP data connection). The types of processing that an IRD performs can vary from creating analog video signals to creating high definition video digital formats. The outputs of an IRD range from simple video outputs to high-speed IP data connections. Companies that produce IRDs commonly offer variations of IRD (such as analog and digital outputs) that meet the specific needs of the CATV system operator.

Figure 1.9 shows the basic function of an integrated receiver device that is used in a cable TV system to receive satellite broadcasted signals and decode the channels. This example shows that an IRD contains a receiver section that can receive and demodulate the MPTS from the satellite. This IRD can decode and decrypt the MPTS to produce several MPEG digital television channels.

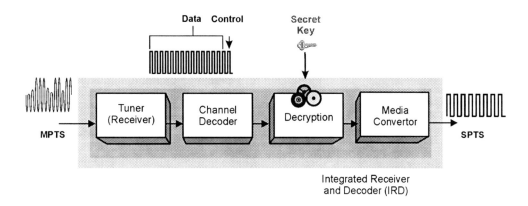

Figure 1.9, Headend Integrated Receiver and Decoder

Off Air Receivers

An off air receiver is a device or assembly that can select, demodulate and decode a broadcasted radio channel. Broadcast receivers are used in cable television systems to receive local broadcasted channels so they can be re-broadcasted in the local cable television system.

In some countries (such as the United States), CATV operators are required to rebroadcast local television channels on their cable television systems. These "must carry" regulations are government requirements that dictate that a broadcaster or other media service provider must retransmit (carry) or make available to carry a type of program or content.

Off-air receivers contain a tuner (receiver), demodulator and decoder for analog and/or digital television signals. The off-air receiver contain a tuning head that allows it to select (or to be programmed to select) a specific television channel. Off-air receivers may be simple analog television tuners (e.g. NTSC, PAL or SECAM) or they may be capable of demodulating and decoding digital television channels (e.g. DTT).

Encoders

An encoder is a device that processes one or more input signals into a specified form for transmission and/or storage. A video encoder is device used to form a single (composite) color signal from a set of component signals. An encoder is used whenever a composite output is required from a source (or recording) that is in a component format.

Transcoders

A transcoder is a device or assembly that enables differently coded transmission systems to be interconnected with little or no loss in functionality.

Rate Shaper

Rate shapers are devices or assemblies in a communication system that adapt and/or transform the transmission rate of one system to the transmission rate of another system. Rate shapers are used as digital turnaround products in a media distribution system (such as a cable television system).

Rate shapers can be used to adjust the data transmission rates for multiple channels that have variable bandwidth rates so they can operate over transmission channel that have constant or limited maximum transmission bit rates. A variable bit rate feed is a media source that has a data transmission rate that varies over time (such as digital video). A constant bit rate feed is a media source that has a data transmission rate that does not vary over time (such as a digital transmission channel).

Figure 1.10 shows how a rate shaper can combine multiple digital television channels to help adjust the maximum bandwidth usage. This example shows 3 digital television channels that have variable bandwidth due to high video activity periods (action scenes with high motion). The combined data rate is shown at the bottom. The combined data rate has a peak data rate that is larger than the 30 Mbps transmission channel can allow. As a

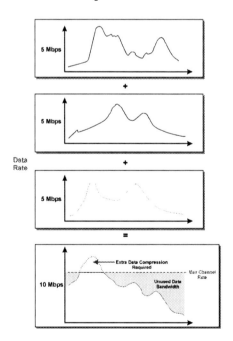

Figure 1.10, MPEG Statistical Multiplexing

result, one or more of the input MPEG channels must use a higher compression rate (temporary lower picture quality).

Channel Modulators

A channel modulator is used to convert video signals into television broadcast channels. Channel modulators are used in cable-TV networks to convert a video program signal (such as CNN or MTV) and converts it with an RF carrier frequency for a television channel that is distributed through the CATV network. The modulator converts both video and audio signals. The frequency of this channel modulator carrier determines the television channel number (e.g., 2 to 120) that the program will be received on by subscribers.

Channel modulators may produce analog or digital signals. Analog modulators take video and audio signals and convert them into a format that can be used by televisions that may be connected to the CATV system (NTSC, PAL or SECAM). Digital modulators convert the digital media programs into a common shared transport stream. Cable television systems commonly use an MPEG transport stream (MPEG-TS) to transfer multiple programs on each RF transmission channel.

MPEG transport streams (MPEG-TS) use a fixed length packet size and a packet identifier identifies each transport packet within the transport stream. A packet identifier in an MPEG system identifies the packetized elementary streams (PES) of a program channel. A program (such as a television show) is usually composed of multiple PES channels (e.g. video and audio).

Channel Processors

A channel processor is a device or assembly that can receive, modify and produce a new channel signal. A common CATV television channel processing function is to change the frequency of the channel signal so that a channel that is received on one frequency can be shifted (translated) to another frequency that is sent on a transmission system.

Channel Signal Combiners

Channel combiners are devices or filter assemblies that allows several modulated carrier signals (physical channels) to be grouped on to the same transmission channel or antenna system.

Each transmission (transport) channel operates on a separate frequency band. CATV systems may allow up to 120 RF channels on a single transmission channel. Because each transport channel can carry multiple logical channels (typically 4 to 6 TV channels per carrier), a digital cable television system can provide hundreds of television channels.

Channel combiners allow multiple RF channels to connect to the same transmission line. To keep signals from the transmitter of one RF channel from being received and interfering with RF transmitters of other transmission channels, the channel combiner provides some port to port isolation (attenuation between the ports).

The types of combiner networks include active and passive combiners. Active combiners include amplifiers to increase the signal level as it passes through the combining network. Passive combiners use filters to isolate the signals from each other.

Figure 1.11 shows how CATV channel combiners are used to combine different RF channel signals into a common transmission line. This diagram shows a passive CATV channel combiner that is composed of multiple signal couplers. Each coupler (tap) allows signals to pass through to the transmission line while providing some isolation (signal attenuation) from other transmitters.

Introduction to Cable Television (CATV), 2nd Edition

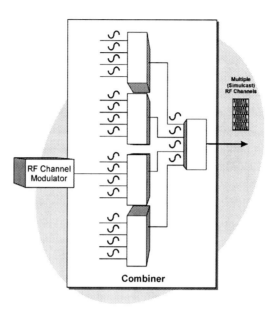

Figure 1.11, CATV Channel Combiners

Asset Management

Asset management is the process of acquiring, maintaining, distributing and the elimination of assets. Assets for television systems are programs or media files. Assets are managed by workflow systems. Workflow management for television systems involves the content acquisition, metadata management, asset storage, playout scheduling, content processing, ad insertion and distribution control.

Content assets are acquired or created. Each asset is given an identification code and descriptive data (metadata) and the licensing usage terms and associated costs are associated with the asset. Assets are transferred into short term or long term storage systems that allow the programs to be retrieved when needed. Schedules (program bookings) are setup to retrieve the assets from storage shortly before they are to be broadcasted to viewers. When programs are broadcasted, they are converted (encoded) into forms that are suitable for transmission (such as on radio broadcast channels or to mobile telephones).

Figure 1.12 shows some of the common steps that occur in workflow management systems. This diagram shows that a workflow management system starts with gathering content and identifying its usage rights. The descriptive metadata for the programs is then managed and the programs are stored in either online (direct), nearline (short delay) or offline (long term) storage systems. Channel and program playout schedules are created and setup. As programs are transferred from storage systems, they may be processed and converted into other formats. Advertising messages may be inserted into the programs. The performance of the system is continually monitored as programs are transmitted through the distribution systems.

Figure 1.12, Television Workflow Management Systems

Content Acquisition

Content acquisition is the gathering content from networks, aggregators and other sources. After content is acquired (or during the content transfer), content is ingested (adapted and stored) into the asset management system.

Ingesting content is a process for which content is acquired (e.g. from a satellite downlink or a data connection) and loaded onto initial video servers (ingest servers). Once content is ingested it can be edited to add commercials, migrated to a playout server or played directly into the transmission chain.

Content acquisition commonly involves applying a complex set of content licensing requirements, restrictions and associated costs to the content. These licensing terms are included in content distribution agreements. Content licensing terms may define the specific type of systems (e.g. cable,

Internet or mobile video), the geographic areas the content may be broadcasted (territories), the types of viewers (residential or commercial) and specific usage limitations (such as number of times a program can be broadcasted in a month). The content acquisition system is linked to a billing system to calculate the royalties and other costs for the media.

Metadata Management

Metadata management is the process of identifying, describing and applying rules to the descriptive portions (metadata) of content assets. Metadata descriptions and formats can vary so metadata may be normalized. Metadata normalization is the adjustment of metadata elements into standard terms and formats to allow for more reliable organization, selection and presentation of program descriptive elements.

Metadata may be used to create or supplement the electronic programming guide (EPG). An EPG is an interface (portal) that allows a customer to preview and select from a possible list of available content media. EPGs can vary from simple program selection to interactive filters that dynamically allow the user to filter through program guides by theme, time period, or other criteria.

Playout Scheduling

Playout scheduling is the process of setting up the event times to transfer media or programs to viewers or distributors of the media. A playout system is an equipment or application that can initiate, manage and terminate the gathering, transferring or streaming of media to users or distributors of the media at a predetermined time schedule or when specific criteria have been met.

Playout systems are used to select and assign programs (digital assets) to time slots on linear television channels. Playout systems are used to setup playlists that can initiate automatic playout of media during scheduled interviews or alert operators to manually setup and start the playout of media programs (e.g. taps or DVDs).

Playout systems may be capable of selecting primary and secondary events. Primary events are the program that will be broadcasted and secondary events are media items that will be combined or used with the primary event. Examples of secondary events include logo insertion, text crawls (scrolling text), voice over (e.g. narrative audio clips) and special effects (such as a squeeze back).

Figure 1.13 shows the playout scheduling involves selecting programs and assigning playout times. This diagram shows a playout system that has multiple linear television channels and that events are setup to gather and playout media programs.

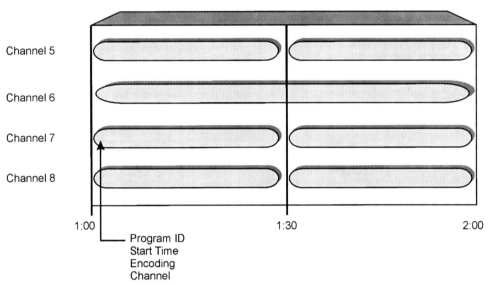

Figure 1.13, Television Playout Scheduling

Because the number of channels and programs is increasing, broadcasters may use playout automation to reduce the effort (workload) to setup playout schedules. Playout automation is the process of using a system that has established rules or procedures that allows for the streaming or transferring media to a user or distributor of the media at a predetermined time, schedule or when specific criteria have been met (such as user registration and payment).

Asset Storage

Asset storage is maintaining of valuable and identifiable data or media (e.g. television program assets) in media storage devices and systems. Asset storage systems may use a combination of analog and digital storage media and these may be directly or indirectly accessible to the asset management system.

Asset management systems commonly use several types of storage devices that have varying access types, storage and transfer capabilities. Analog television storage systems may include tape cartridge (magnetic tape) storage. Digital storage systems include magnetic tape, removable and fixed disks and electronic memory.

Asset storage devices are commonly setup in a hierarchical structure to enable the coordination of storage media. Some of the different types of storage systems include cache storage (high speed immediate access), online storage, nearline storage, and offline storage.

Online storage is a device or system that stores data that is directly and immediately accessible by other devices or systems. Online storage types can vary from disk drives to electronic memory modules. Media may be moved from one type of online storage system to another type of online storage system (such as a disk drive) to another type of online storage (such as electronic memory) that would allow for rapid access and caching. Caching is a process by which information is moved to a temporary storage area to assist in the processing or future transfer of information to other parts of a processor or system.

Nearline storage is a device or system that stores data or information that is accessible with some connection setup processes and/or delays. The requirement to find and/or setup a connection to media or information on a nearline storage system is relatively small. Data or media that is scheduled to be transmitted (e.g. broadcasted) may be moved to nearline storage before it is moved to an online storage system.

Offline storage is a device or system that stores data or information that is not immediately accessible. Media in offline storage systems must be located and setup for connection or transfer to be obtained. Examples of offline storage systems include storage tapes and removable disks.

Content Processing

Content processing is adaptation, modification or merging of media into other formats. Content processing may include graphics processing, encoding and/or transcoding.

A graphics processor is an information-processing device that is dedicated for the acquisition, analysis and manipulation of graphics images. Graphics processing may be required to integrate (merge or overlay) graphic images with the underlying programs.

Content encoding is the manipulation (coding) of information or data into another form. Content encoding may include media compression (reducing bandwidth), transmission coding (adapting for the transmission channel) and channel coding (adding control commands for specific channels).

Transcoding is the conversion of digital signals from one coding format to another. An example of transcoding is the conversion of MPEG-2 compressed signals into MPEG-4 AVC coded signals.

Ad Insertion

Ad insertion is the process of inserting an advertising message into a media stream such as a television program. For broadcasting systems, Ad inserts are typically inserted on a national or geographic basis that is determined by the distribution network. For IP television systems, Ad inserts can be directed to specific users based on the viewer's profile.

An advertising splicer is a device that selects from two or more media program inputs to produce one media output. Ad splicers receive cueing messages (get ready) and splice commands (switch now) to identify when and which media programs will be spliced.

Cue tones are signals that are embedded within media or sent along with the media that indicates an action or event is about to happen. Cue tones can be a simple event signal or they can contain additional information about the event that is about to occur. An example of a cue tone is a signal in a television program that indicates that a time period for a commercial will occur and how long the time period will last.

Analog cue tone is an audio sequence (such as DTMF tones) that indicates a time period that will be available ("avail") for the insertion of another media program (e.g. a commercial).

An 'avail' is the time slot within which an advertisement is placed. Avail time periods usually are available in standard lengths of 10, 20, 30 or 40 seconds each. Through the use of addressable advertising, which may provide access to hundreds of thousands of ads with different time lengths, it is possible for many different advertisements, going to different audiences to share a single avail.

Digital program insertion is the process of splicing media segments or programs together. Because digital media is typically composed of key frames and difference pictures that compose a group of pictures (GOP), the splicing of digital media is more complex than the splicing of analog media that has picture information in each frame which allows direct frame to frame splicing.

Figure 1.14 shows how an ad insertion system works in a CATV network. This diagram shows that the program media is received and a cue tone indicates the beginning of an advertising spot. When the incoming media is received by the splicer/remultiplexer, it informs the ad server that an advertising media clip is required. The ad server provides this media to the splicer which splices (attaches) each ad to the appropriate media stream. The resulting media stream with the new ad is sent to the viewers in the distribution system.

Introduction to Cable Television (CATV), 2nd Edition

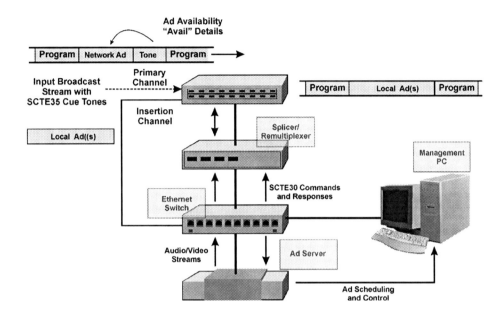

Figure 1.14, Television Ad Splicer

Distribution Control

Distribution control is the processes that are used to route products or service uses to get from the manufacturer or supplier to the customer or end user. Media broadcasters may have several types of distribution networks including radio broadcast systems, cable television distribution, mobile video and Internet streaming.

Distribution systems use a mix of media encoding formats that can include MPEG, VC-1 along with other compressed forms. The transmission of media to viewers ranges from broadcast (one to many), multicast (point to multipoint) and unicast (point to point).

Asset management systems use work orders to define, setup and manage assets. A work order is a record that contains information that defines and quantifies a process that is used in the production of media (e.g. television programs) or services. The development and management of assets is called workflow.

As the number of available programs and channels increases, it is desirable to automate the workflow process. Workflow automation is the process of using a system that has established rules or procedures that allows for the acquisition, creation, scheduling or transmission of content assets.

Distribution Network

The distribution network is the part of a cable television system that connects the head-end of the system (video and media sources) to the customer's equipment. Traditionally, the local connection has been composed of a coaxial cable that allows for the one-way transmission of a maximum of one hundred and twenty 6 MHz RF television signals or approximately one hundred 8 MHz RF channels.

Core Network

The core network is the central network portion of a communication system. The core network primarily provides interconnection and transfer between edge networks. Core networks in CATV systems are commonly setup as fiber rings and spurs. A fiber ring is an optical network of network topology with a connection that provides a complete loop. The ring topology is used to provide a backup distribution path as traffic to be quickly rerouted in the other direction around the loop in the event of a fiber cut. A fiber spur is a fiber line that extends the fiber ring into another area for final distribution.

CATV systems commonly use a mix of fiber rings in the core and coax lines (hybrid fiber and coax) to connect the customer. The hybrid fiber coax (HFC) system provides high-speed backbone data interconnection lines (the fiber portion) to interconnect end user video and data equipment. HFC systems convert (shift) the RF channels at the head end into optical signal that can travel down a fiber. When the optical signal reaches a fiber node, it is converted (downshifted) back onto the radio frequency band which then travels down the coaxial line.

Figure 1.15 shows a typical cable distribution system that uses a combination of fiber optic cable for the core distribution and coaxial cable for the local connection. This diagram shows that the multiple RF television channels at the head-end of the cable television system are shifted in frequency to allow distribution through high-speed fiber cable. The fiber cable is connected in a loop around the cable television service area so that if a break in the cable occurs, the signal will automatically be available from the other part of the loop. The loop is connected (tapped) at regular points by a fiber hub that can distribute the optical signals on fiber spurs. The fiber spurs end into fiber nodes that convert the optical signals into RF television signals that are distributed on the local coaxial cable network.

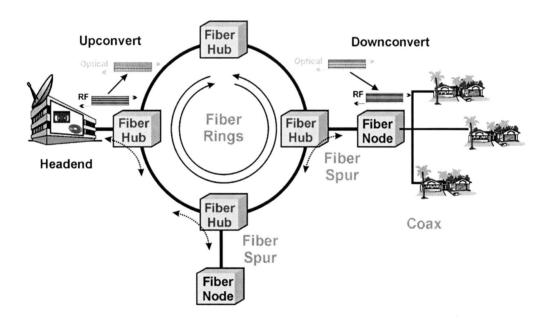

Figure 1.15, Hybrid Cable Television Distribution Network

Access Network

Access network is a portion of a communication network that allows individual subscribers or devices to connect to the core network. For CATV systems, the end user device (e.g. set top box) communicates to the system using RF channels. There are several types of RF channels that may be used in CATV systems and these include analog channels, digital channels, out of band control channels and data channels.

Analog RF channels transfer television programming in analog form (e.g. NTSC or PAL). These signals may be available for anyone to use or they may be scrambled so the receiver must decode (descramble) the analog program.

Digital channels transfer television programming in digital form (e.g. MPEG). The programs transferred by digital channels are usually compressed so each digital RF channel carries multiple television programs. Digital channels may be available for anyone to use or they may be encrypted so the receiver must decode (decrypt) the digital program.

Cable systems send (and optionally receive) control information to the devices such as channel identification information and programming guides. The control messages may be sent on the television channel that is displaying the program (In band) or on separate control channels (out of band). When the data is sent in band, the control data shares the bandwidth with the television programming. When the data is sent on out of band channels, it is sent on other RF channels.

The channel information is typically repeated in carousel form so that receivers can capture and store the information when it becomes available. If the receiver cannot obtain the entire data block of information, it can simply wait until the next transmission of data. Each unique block of information is assigned an identification code to allow the receiver to determine if it has already received the block or if it is new block of information it needs to decode and store.

Data RF channels are designed to efficiently transfer user data (such as Internet data) between users and the cable system. Data RF channels can use a very efficient form of modulation (such as QAM) from the system to the end user allowing the cable system to provide high speed data from the system to the receiver (up to 30 Mbps to 40 Mbps) per RF channel. Because signals from multiple users are combined when they are sending data to the cable system, this increases the amount of noise level so a more robust (less efficient) modulation form is used (such as QPSK) which can provide medium speed data from the users to the system (up to 2 to 5 Mbps) per RF channel. Data channels are defined in the DOCSIS specification available from CableLabs.

Because data channels can transfer information much faster than the older out of band (OOB) channels, when data channels are available in a system, the data channel can also send the system information so the out of band channels do not need to be used. However, systems that upgrade to data channels may continue to use the OOB channels as the existing customer equipment may not have data channel capability.

User devices (such as STBs) may only have the capability to receive one type of RF channel at a time (single tuner) or it may be able to simultaneously receive multiple channels at the same time (multiple tuners). For cable receivers that only have one tuner, the user may be interrupted when data (e.g. program guides) is gathered. For cable receivers that have multiple tuners, the control information can be received and sent without interrupting the viewer's program display.

Figure 1.16 shows how the access portion of a CATV uses RF channels to communicate with a set top box. This example shows a CATV network that has a mix of analog TV, digital TV, control channels and data communication channels. System information may be repeatedly sent using in band, out of band or on data channels. This example shows that the CATV system may communicate with older set top boxes (STBs) that only have one tuner or it may be communicating with set top boxes that have multiple tuners. The STBs with multiple tuners can simultaneously receive programs and system information that allows the viewer to continue to watch their programs without interruption.

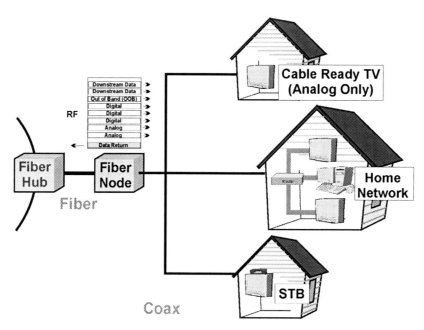

Figure 1.16, Cable Television Access Network

Premises Distribution

A premises distribution network is the equipment and software that is used to transfer data and other media in a customer's facility, home or personal area. A PDN is used to connect terminals (computers) to other networks and to wide area network connections. Some of the common types of PDN are wired Ethernet, Wireless LAN, Powerline, Coaxial and Phoneline Data.

PDN networking systems have transitioned from low speed data, simple command and control systems to high-speed multimedia networks along with the ability to transfer a variety of media types that have different transmission and management requirements. Each of the applications that operate through a PDN can have different communication requirements that typically includes a maximum data transmission rate, continuous or bursty transmission, packet delay and jitter and error tolerance. The PDN system may manage these connections using a mix of protocols that can define and manage quality of service (QoS). Transmission medium types for premises distribution include wired Ethernet (data cable), wireless, powerline, phoneline and coaxial cables.

Wired LAN systems use cables to connect routers and communication devices. These cables can be composed of twisted pairs of wires or optical fibers. Wired LAN data transmission rates vary from 10 Mbps to more than 1 Gbps. While wired Ethernet systems offer high data throughput and reliability, many homes do not have dedicated wiring installed for Ethernet LAN networks and for the homes that do have data networks, the data outlets are often located near computers rather than near televisions.

Wireless local area network (WLAN) systems allow computers and workstations to communicate with each other using radio propagation as the transmission medium. The wireless LAN can be connected to an existing wired LAN as an extension, or it can form the basis of a new network. Wi-Fi television distribution is important because it is an easy and efficient way to get digital multimedia information where you need it without adding new wires. Wireless LAN data transmission rates vary from 2 Mbps to over 54 Mbps and higher data transmission rates are possible through the use of channel bonding. WLAN networks were not designed specifically for multimedia. In the mid 2000s, several new WLAN standards were created to enable and ensure different types of quality of service (QoS) over WLAN.

Power line carrier systems allow signals to be simultaneously transmitted on electrical power lines. A power line carrier signal is transmitted above the standard 60 Hz power line power frequency (50 Hz in Europe). Power line premises distribution for television is important because televisions, set-top boxes, digital media adapters (DMAs) and other media devices are already connected to power outlets already installed in a home or small businesses. Older (legacy) power line communication systems had challenges with wiring systems that used two or more phases of electrical power. Today, with the benefit of modern signal processing techniques and algorithms, most of these impairments no longer are an impediment to performance and some powerline data systems have data transmission rates of over 200 Mbps.

Coaxial cable premises distribution systems transfer user information over coaxial television lines in a home or building. Coaxial distribution systems may simply distribute (split) the signal to other televisions in the home or they may be more sophisticated home data networks. When coax systems are setup as simple distribution systems, they are setup as a tree distribution system. The root of the tree is usually at the entrance point to the home or building. The tree may divide several times as it progresses from the root to each television outlet through the use of signal splitters. When coaxial systems are setup as data networks, data signals at high frequencies (above 860 MHz) are combined with broadcast signals over the same coaxial lines. Coaxial cable data transmission rates vary from 1 Mbps to over 1 Gbps and many homes have existing cable television networks and the outlets which are located near video accessory and television viewing points.

Telephone wiring premises distribution systems transfer user information over existing telephone lines in a home or building. Because telephone lines may contain analog voice signals and data signals (e.g. DSL), premises distribution on telephone lines uses frequency bands above 1 MHz to avoid interference with existing telephone line signals. Telephone data transmission rates vary from 1 Mbps to over 300 Mbps. There are typically several telephone line outlets installed in a home and they may be located near television viewing points, making it easy to connect television-viewing devices.

Figure 1.18 shows how home coaxial cable lines can be used to distribute data and television signals. This diagram shows that coax lines to and from the cable television (CATV) company may contain analog and digital television and modem signals. Cable television distribution systems use lower frequencies for uplink data signals and upper frequencies for downlink data and digital television signals. Some of the center frequencies are used for analog television signals. These frequency bands typically range up to 1 GHz. Coax premises distribution systems use frequencies above the 1 GHz frequency band to transfer signals to cable television jacks throughout the house. Adapter boxes or integrated communication circuits convert the video and/or data signals to high frequency channels that are distributed to different devices located throughout the house. To ensure the coax premises distribution signals do not transfer out of the home to other nearby homes, a blocking filter may be installed.

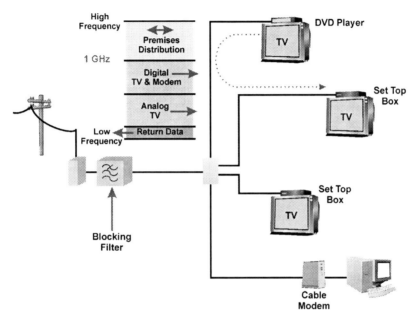

Figure 1.18, Cable Television Premises Distribution

CATV End User Devices

Cable television end user devices adapt RF channels on a cable connection to a format that is accessible by the end user. Cable television user devices are commonly located in a customer's home to allow the reception of video signals on a television. The key functions for end user devices include interfacing to the cable network, selecting and decoding channels, processing the media into a form usable by humans and providing controls that allow the user to interact with the device.

The network interface for cable television end user devices allow it to receive analog broadcast, digital broadcast, digital control and data channels. The device must select the appropriate RF channel and separate out the component parts of the television signal (video, audio and data). The underlying media is then decoded and decrypted (unscrambled). The media is then converted (rendered) into a form that can be displayed or heard by the end user. A program guide and menu system is provided to allow the user to navigate and select features and services.

Figure 1.19 shows the basic functions of a cable television end user viewing device. This diagram shows that the end user device has a network interface, signal processing, decoding, rendering and user interface. The network interface may contain one or several RF tuners to receive and decode broadcast and control channels. The signal processing receives, selects and demultiplexes the incoming channels. After the channels are received, the channel may require decoding (decryption) for scrambled channels. The STB then converts the data into signals that can be displayed to the viewer (rendering). The STB has a user interface, which allows the system to present information to the user (such as the program guide) and to allow the user to interact (select channels) with the STB.

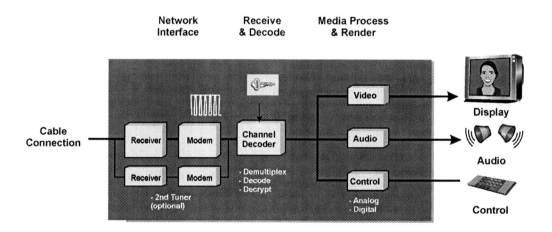

Figure 1.19, IPTV User Device Functions

Cable access device capabilities include number of tuners, display capability, media processing (video, audio and graphics), security, software applications, accessories, middleware compatibility, media distribution and upgradeability.

Tuners

A tuner is the radio frequency and intermediate frequency parts of a radio receiver that produce a low-level audio output signal. Set top boxes may have one or more tuners. Set top boxes with multi-channel tuners are capable of simultaneously receiving two or more communication channels.

Display Capability

Display capability is the ability of a device to render images into a display area in different formats. Display capabilities for STB include size and resolution (SD or HD), the type of video (interlaced or progressive) and display positioning (scaling and displaying multiple sources). Display capabilities for television systems are characterized in the MPEG industry standards and the sets of capabilities (size and resolution) are defined as MPEG profiles.

Security

Security for set top boxes is the ability to maintain its normal operation without damage and to ensure content that is accessed by the STB is not copied or used in an unauthorized way by the user. Set top boxes commonly include smart cards and security software to ensure the content is used in its authorized form.

A smart card is a portable credit card size device that can store and process information that is unique to the owner or manager of the smart card. When the card is inserted into a smart card socket, electrical pads on the card connect it to transfer information between the electronic device and the card. Smart cards are used with devices such as mobile phones, television set top boxes or bank card machines. Smart cards can be used to identify and validate the user or a service. They can also be used as storage devices to hold media such as messages and pictures.

Smart card software can be embedded (included) in the set top box to form a virtual smart card. A virtual smart card is a software program and associated secret information on a users device (such as a TV set top box) that can store and process information from another device (a host) to provide access control and decrypt/encrypt information that is sent to and/or from the device.

A digital rights management client is a computer, hardware device or software program that is configured to request DRM services from a network. An example of a DRM client is a software program (module) that is installed (loaded) into a converter box (e.g. set top box) that can request and validate information between the system and the device in which the software is installed.

A secure microprocessor is a processing device (such as an integrated circuit) that contains the processes that are necessary to encrypt and decrypt media. Secure microprocessors contain the cryptographic algorithms such as DES, AES or PKI. The secure microprocessor can be a separate device or it can be a processing module that is located within another computing device (such as a DSP).

Media Processing

Media processing is the operations used to transfer, store or manipulate media (voice, data or video). The processing of media ranges from the playback of voice messages to modifying video images to wrap around graphic objects (video warping). Media processing in set top boxes includes video processing, audio processing and graphics processing.

Video processing is the methods that are used to convert and/or modify video signals from one format into another form using signal processing. An example of video processing is the decoding of MPEG video and the conversion of the video into a format that can be displayed on a television monitor (e.g. PAL or NTSC video).

Audio processing is the methods that are used to convert and/or modify audio signals from one format into another form using signal processing. An example of audio processing is the decoding of compressed audio (MP3 or AAC) and conversion into multiple channels of surround sound audio (5.1 audio).

Graphics processing is the methods that are used to convert and/or modify image objects from one format into another form. An example of graphics processing the conversion of text (e.g. subtitles) into bitmapped images that can be presented onto a television display (on screen display).

Software Applications

A software application is a software program that performs specific operations to enable a user to apply the software to their specific needs or problems. Software applications in set top boxes may be in the form of embedded applications, downloaded applications or virtual applications.

Embedded applications are programs that are stored (encapsulated) within a device. An example of an embedded application is a navigation browser that is included as part of a television set top box. Downloaded applications are software programs that are requested and transferred from the system when needed. A loader application (the loader is an embedded application) is used to request and transfer applications from the system. Virtual applications are software instructions that are written in another language to perform application using an interpreter program (e.g. Javascript).

Accessories

Cable television accessories are devices or software programs that are used with cable systems or services. Examples of cable television accessories include remote controls, gaming controllers and other human interface devices that are specifically designed to be used with cable television systems and services. These accessories may have dedicated connection points (such as game controllers) or they may share a standard USB connection.

Middleware Compatibility

Middleware compatibility is the ability of a device to accept software programs (clients) that interface the device to other software and operates between the system host (servers) and the end user interface (clients). A middleware client is a software module that is installed in a device that is configured to request and deliver media or services from a server (e.g. to request television programs from media network).

Upgradability

Upgradability is the ability of a device or system to be modified, changed or use newer components and/or technology innovations as they become available. The ability to upgrade the capabilities of a set top box may be performed by software downloads or through the use of software plug-ins.

A plug-in is a software program that works with another software application to enhance its capabilities. An example of a plug-in is a media player for a web browser application. The media player decodes and reformats the incoming media so it can be displayed on the web browser.

Media Portability

Media portability is the ability to transfer media from one device to another. Media portability can range from stored media locally in a hard disk (for personal video recorder) or by shared media through home connections (such as a premises distribution network).

CATV Access Devices

An access device is a conversion assembly that receives a transmitted signal and makes it perceptible to a human user or converts it into some other useful form. CATV access devices include set top boxes, cable ready televisions, cable modems and cable telephones.

Cable Ready Televisions

Cable ready television is a video display device (a television) that is capable of receiving and displaying channels from a cable television system without the need for external adapters or devices. Cable ready televisions can be analog or digital cable ready.

To be analog cable ready, the tuner or receiver needs to be capable of adjusting its frequency and demodulate cable television channels. To be digital cable ready, the television has a Cablecard slot so the television can decode encrypted channels. Digital cable ready televisions may have RF connections or Ethernet only (e.g. for hotels).

Set-Top box (Cable Converters)

Cable converters, commonly called a "set-top" box are electronic devices that convert an incoming cable television signal into a form that can be displayed on a video device, typically a television or computer. The set-top box is typically located in a customer's home to enable the reception and/or interaction with services on the customer's television or computer. In digital cable systems, a set-top box is also used to convert digital video (e.g., MPEG2) into standard NTSC or PAL video formats that is used for standard televisions.

Figure 1.20 shows a basic hardware architecture diagram for a cable set top box (STB). This diagram shows that a cable STB is composed of tuner/receiver components, a microprocessor (uP), memory components, media processors, audio/visual interfaces and user interface controls. The tuner/receiver components adapt the physical transmission formats from cable networks (analog or digital RF channels) into a format that can be processed by the cable STB. This diagram shows that the cable STB may contain multiple tuners to allow the STB to receive television programming while it is receiving information from other channels (such as the television programming guide). The microprocessor controls the overall operation of the STB. The media processor is a special purpose digital signal processor that can convert and manipulate the media (such as converting MPEG). Cable STBs have multiple types of memory that range from short term random access

Introduction to Cable Television (CATV), 2nd Edition

memory that enables the uP to process instructions to unchangeable read only memory that holds the operating system instructions. The audio/visual interfaces adapt the media into formats that can be displayed or heard by the user. The user interface contains displays, keypads and remote control interfaces to allow the user to interact with the STB.

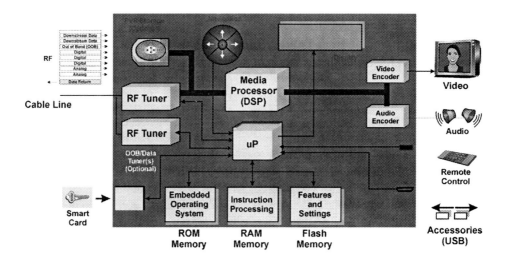

Figure 1.20, Cable Set top Box Hardware

Some cable set top boxes have the capability of receiving and processing signals from other broadcast systems. These hybrid set top boxes (HSTB) are an electronic device that adapts multiple types of communications mediums (RF or data signals) to a format that is accessible by the end user. The use of HSTBs allows a viewer to get direct access to broadcast content from other systems such as satellite systems, DTT and/or interactive IPTV via a broadband network.

Cable Modems

A cable modem converts RF signals from the cable system into a standard data format that can be used by computers and converts data signals from a computer into a form that can be routed back to the data network. Cable modems select and decode high data-rate signals on the cable television system (CATV) into digital signals that are designated for a specific user.

There are two generations of cable modems; First Generation one-way cable modems transmit high speed data to all the users into a portion of a cable network and return low speed data through telephone lines or via a shared channel on the CATV system. Second generation cable modems offer data transmission rates in both downstream and upstream directions. Second generation cable television systems use two-way fiber optic cable for the head end and feeder distribution systems. This allows a much higher data transmission rate and many more channels available for each cable modem.

Cable television systems commonly use some of the upper RF channels for downstream data channel and lower frequency RF channels are used for upstream. The downstream channels can use very efficient QAM modulation which offer data rates of 30 Mbps to 40 Mbps for each RF channel and a more robust QPSK modulation used on the upstream for data rates of approximately 2 Mbps to 4 Mbps.

Figure 1.21 shows a block diagram of a cable modem. This diagram shows that a cable modem has a tuner to convert an incoming 6 MHz or 8 MHz RF channel to a low frequency baseband signal. This signal is demodulated to a digital format, demultiplexed (separated) from other digital channels, and is decompressed to a single data signal. This data signal is connected to a computer typically in Ethernet data format (e.g. 10 Base T Ethernet). Data that is sent to the modem is converted to either audio signals for transfer via a telephone line (hybrid system) or converted to an RF signal for transmission back through the cable network.

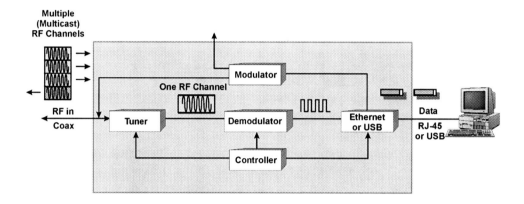

Figure 1.21, Cable Modem

Cable Telephone Adapters

Cable telephone adapters are devices that convert telephone signals into another format (such as digital Internet protocol) that can be transferred on a cable television system. These adapter boxes may provide a single function such as providing digital telephone service or they may convert digital signals into several different forms such as audio, data, and video. When adapter boxes convert into multiple information forms, they may be called multimedia terminal adapters (MTAs) or integrated access devices (IADs).

Cable telephone adapters must convert both the audio signals (voice) and control signals (such as touch-tone or hold requests) into forms that can be sent and received via the cable television network.

Market Growth

In 2006, of the 1.2 billion TV households worldwide, 355 million are cable TV households. [8]. The key segments of the cable television market include cable television, cable modem and cable telephony. Overall, the growth of television subscribers is moderate (2% to 3% yearly) while the growth of advanced services such as cable modem (broadband Internet) and cable telephone services are high (20% to 40% yearly).

Cable Television Market

Cable television service market was 364 million households in 2007. The largest countries with cable television subscribers are 106 million TV households in China and 69 million in USA.

The largest penetration of cable services is the United States. The first television sets in the United States were put into operation in 1946. By 1949, receiver sales were exceeding 10,000 per month. By 2000, over 98% of all American households had at least one TV and more than 70% of American households owning two or more television sets. Cable television service started in the United States during the 1960s.

Figure 1.22 shows that the worldwide cable television market has steadily grown to over 355 million customers by 2007. This diagram shows the growth rate of new cable television subscribers has been 2% to 3% each year.

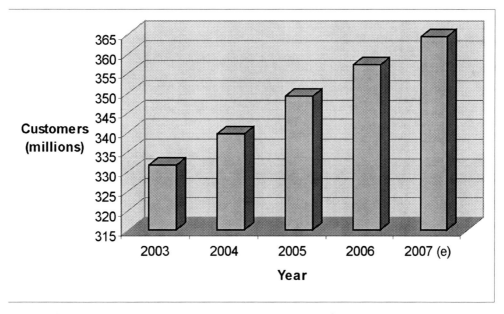

Figure 1.22, Cable Worldwide Cable Television Market Growth 2003 to 2007
Source: In-Stat

Cable Modem Market

Cable modem service market was 71 million households in 2007. Cable modem services revenues increased from $22 billion in 2005 to $26 billion in 2006. The broadband cable services market has aggressive competition from DSL (outside the USA) and competition is likely to increase with wireless broadband (e.g. WiMax) and high speed mobile communication (e.g. 3G cellular).

The demand for cable modems is part of the demand for broadband connectivity. Global projections show that over 600 million households will have broadband connections by 2010 and the value of the global broadband market will exceed $580 billion over the next 10 years [9].

Figure 1.23 shows that the worldwide cable modem market has steadily grown to over 71 million customers by 2007. This diagram shows the while the growth of new cable modem subscribers is high, the growth rate has decreased from 28% in 2003 to 18% in 2007.

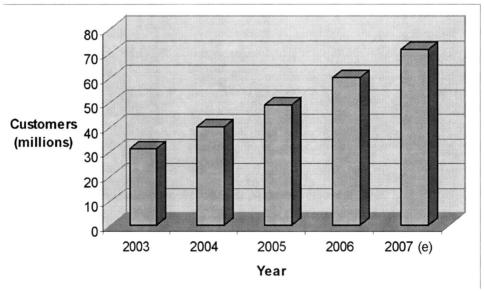

Figure 1.23, Worldwide Cable Modem Market 2003 to 2006
Source: In-Stat

Cable Telephone Market

The cable telephone service market was 29 million households worldwide in 2007. Cable telephone service is transitioning from basic telephone service to offer advanced fixed and mobile telephone services.

The market penetration for cable television customers who select telephone services is less than 8% so there is significant potential for the continued growth of the cable telephony market. The growth of the cable telephone market is increasing at 20% to 40% per year.

Figure 1.24 shows that the worldwide cable telephone market has steadily grown to over 29 million customers by 2007. This diagram shows the while the growth of new cable telephone subscribers remains high, the growth rate increased from 22% in 2003 to 41% in 2006 but in 2007 it is expected to grow at 30%.

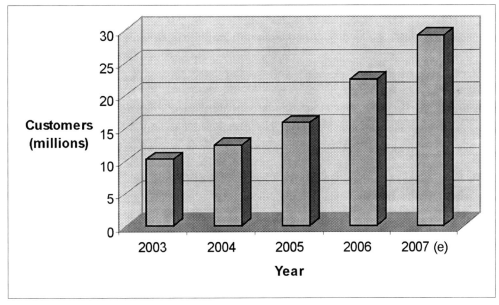

Figure 1.24, Worldwide Cable Telephone Market Growth 2003 to 2007
Source: In-Stat

Technologies

Some of the key technologies used in CATV systems include analog video, digital video, signal scrambling, cable modem, and high definition television (HDTV).

Analog Video

Analog video contains a rapidly changing signal (analog) that represents the luminance and color information of a video picture. Sending a video picture involves the creation and transfer of a sequence of individual still pictures called frames. Each frame is divided into horizontal and vertical lines. To create a single frame picture on a television set, the frame is drawn line by line. The process of drawing these lines on the screen is called scanning. The frames are drawn to the screen in two separate scans. The first scan draws half of the picture and the second scan draws between the lines of the first scan. This scanning method is called interlacing. Each line is divided into pixels that are the smallest possible parts of the picture. The number of pixels that can be displayed determines the resolution (quality) of the video signal. The video signal breaks down the television picture into three parts: the picture brightness (luminance), the color (chrominance), and the audio.

There are three primary systems used for analog television broadcasting: NTSC, PAL, and SECAM. The National Television System Committee (NTSC) is used for the Americas, while PAL and SECAM are primarily used in the UK and other countries. The major difference between the analog television systems is the number of lines of resolution and the methods used for color transmission.

There have been enhancements made to analog video systems over the past 50 years. These include color video, stereo audio, separate audio programming channels, slow data rate digital transfer (for closed captioning) and ghost canceling. The next major change to television technology will be its conversion to HDTV.

Figure 1.25 demonstrates the operation of the basic analog television system. The video source is broken into 30 frames per second and is converted into multiple lines per frame. Each video line transmission begins with a burst pulse (called a sync pulse) that is followed by a signal that represents color and intensity. The time relative to the starting sync is the position on the line from left to right. Each line is sent until a frame is complete and the next frame can begin. The television receiver decodes the video signal to position and control the intensity of an electronic beam that scans the phosphorus tube ("picture tube") to recreate the display.

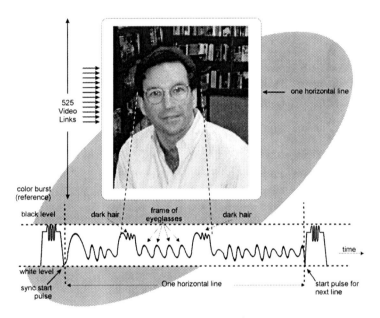

Figure 1.25, Analog Video

Digital Video

Digital broadcasting is the sending of a digital signal through a common channel to a group of users that may be capable of decoding some or all of the broadcast information.

Digital Television (DTV) is a method of transferring video images and their audio components through digital transmission. There are several formats used for DTV including high quality digital MPEG and 28.8 video.

Digital video is the sending of a sequence of picture signals (frames) that are represented by binary data (bits) that describe a finite set of color and luminance levels. Sending a digital video picture involves the conversion of a scanned image to digital information that is transferred to a digital video receiver. The digital information contains characteristics of the video signal and the position of the image (bit location) that will be displayed. Digital television continues to send information in the form of frames and pixels. The major difference is the frames and pixels are represented by digital information instead of a continuously varying analog signal.

The first digital television broadcast license for the United States was issued to a Hawaiian television station in September 1997. Digital television sends the video signal in digital modulated form. Ironically, many television signals have been captured and stored in digital form for over 10 years. To transmit these digital video signals, they must first be converted to standard analog television (NTSC or PAL) to be transmitted through analog transmission systems and to reach analog televisions.

When digital transmission is used, most digital video systems use some form of data compression. Data compression involves the characterization of a single picture into its components. For example, if the picture was a view of the blue sky, this could be characterized by a small number of data bits that indicate the color (blue) and the starting corner and ending corner. This may require fewer than 10 bytes of information. When this digital information is received, it will create a blue box that may contain over 7,200 pixels. With a color picture, this would have required several thousand bytes of information for only 1 picture.

In addition to the data compression used on one picture (one frame), digital compression allows the comparison between frames. This allows the repeating of sections of a previous frame. For example, a single frame may be a picture of a city with many buildings. This is a very complex picture and data compression will not be able to be as efficient as the blue sky example above. However, the next frame will be another picture of the city with only a few changes. The data compression can send only the data that has changed between frames.

Digital television broadcasting that uses video compression technology allows for "multicasting" (simultaneously sending) several "standard definition" television channels (normally up to five channels) in the same bandwidth as a standard analog television channel. Unfortunately, high definition digital television channels require a much higher data transmission rate and it is likely that only a single HDTV channels can be sent on a digital television channel.

Figure 1.26 demonstrates the operation of the basic digital video compression system. Each video frame is digitized and then sent for digital compression. The digital compression process creates a sequence frames (images) that start with a key frame. The key frame is digitized and used as reference points for the compression process. Between the key frames, only the differences in images are transmitted. This dramatically reduces the data transmission rate to represent a digital video signal as an uncompressed digital video signal requires over 50 Mbps compared to less than 4 Mbps for a typical digital video disk (DVD) digital video signal.

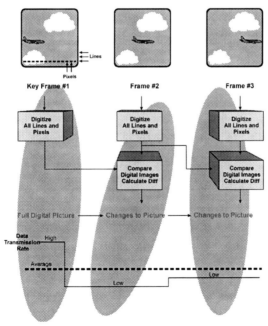

Figure 1.26, Digital Video

Rights Management

Rights management is a process of organization, access control and assignment of authorized uses (rights) of content. Rights management may involve the control of physical access to information, identity validation (authentication), service authorization, and media protection (encryption). Rights management systems are typically incorporated or integrated with other systems such as content management system, billing systems, and royalty management and they use signal scrambling and/or encryption to protect premium content.

Video signal scrambling is a deliberate act of changing an electrical signal (often distortion of video, digital voice, or data) to prevent interpretation of the signals by unauthorized users that are able to receive the signal. Because the scrambling process is performed according to a known procedure or algorithm, the received signal can be descrambled to recover the original digital stream through the use of a known code or filtering technique.

In 1971, the first system using scrambling on a cable system was demonstrated [10]. The first scrambling suppressed the synchronization signal so the video of the television picture was distorted. To decode the scrambled video, the synchronization signal was recreated in the setup box by decoding the correct synchronization signal from another portion of the transmitted signal. Another form of signal scrambling that was used was the insertion a signal that was slightly offset from the channel's frequency to interfere with the picture.

These early video signal scrambling systems were relatively simple in design. As a result, accessory devices soon became available that allowed consumers to decode the scrambled signals without paying subscription fees. To prevent unauthorized viewing, more sophisticated signal scrambling technologies have been developed.

For digital television signals, video signals can be easily encrypted with a key code. To successfully decode the video signal, the set-top box must contain the decryption key code. For two-way cable systems, this code can be dynamically changed and unauthorized viewing has been greatly reduced.

Cable Modems

A cable modem is a device that MOdulates/DEModulates data signals on a coaxial cable and divides the high data rate signals into digital signals designated for a specific user. Cable modems are often asymmetrical modems as the data transfer rate in the downstream direction is typically much higher than the data transfer in the upstream direction. The typical gross (system) downstream data rates range between 30-40 Mbps and gross upstream data rates typically range up to 10 Mbps. New versions of cable data systems can combine channels in either direction providing for data transmission rates of over 160 Mbps.

Cable modems contain a tuner, a demodulator, a modulator, media access control (MAC) section, and a control section. The tuner converts a selected RF channel (high frequency) to the modem baseband (low frequency) signal. The tuner makes adjustments to a frequency (usually between 42 and 850 MHz) for downstream traffic and may convert the upstream traffic to a different RF channel (usually between 5 and 42 MHz).

Early cable modems used a hybrid system that used the cable system for downlink channels and a telephone line for upstream traffic. This was desirable as many of the amplifiers in the cable television system only provided for one-way amplification.

The cable modem receiver contains a demodulator that converts the low frequency received signal into its original baseband digital form and performs error detection and correction. The cable modem may contain a decoder to convert compressed video into a form that can be displayed on the computer monitor. The modulator converts the digital information from the computer into a format suitable for transfer back to the Internet. For hybrid systems, this may be a telephone line audio modem and for two-way cable systems, the modulator converts the data into radio-frequency signals for transmission on the cable system. A control section coordinates the upstream and downstream access operations (called media access control - MAC) of the cable modem. The control section also coordinates the overall operation of the cable modem including how it interfaces to communication devices. For example, the data may be converted to Ethernet format for communication with a personal computer.

Figure 1.27 shows a basic cable modem system that consists of a head end (television receivers and cable modem system), distribution lines with amplifiers, and cable modems that connect to customers' computers. This diagram shows that the cable television operator's head end system contains both analog and digital television channel transmitters that are connected to customers through the distribution lines. The distribution lines (fiber and/or coaxial cable) carry over 100 television RF channels. Some of the upper television RF channels are used for digital broadcast channels that transmit data to customers and the lower frequency channels are used to transmit digital information from the customer to the cable operator. Each of the upper digital channels can transfer 30 to 40 Mbps and each of the lower digital channels can transfer data at approximately 2 to 10 Mbps. The cable operator has replaced its one-way distribution amplifiers with precision (linear) high frequency bi-directional (two-way) amplifiers. Each high-speed Internet customer has a cable modem that can communicate with the cable modem termination system (CMTS) modem at the head end of the system where the CMTS system is connected to the Internet.

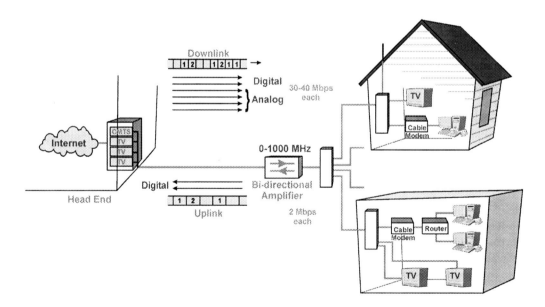

Figure 1.27, Cable Modem System

High Definition Television (HDTV)

High definition television (HDTV) is a TV broadcast system that proves higher picture resolution (detail and fidelity) than is provided by conventional NTSC and PAL television signals. HDTV signals can be in analog or digital form.

HDTV has been offered in several countries since its introduction in Japan in 1988. The first HDTV receivers in the United States were introduced at the 1998 Winter Consumer Electronics Show in Las Vegas.

HDTV radio broadcast channels can use the same 6 MHz channel bandwidth. However, it must replace the existing NTSC signal with a new high resolution analog or high speed digital radio signal. Initial demonstrations of HDTV required 2 standard television channels. The FCC has finally approved the "Grand Alliance" standard for high-definition television for the United States that only requires one standard television channel to send a HDTV digital channel and supplementary services.

The FCC introduced a new table of digital television channel numbers and RF power level assignments for existing full-power television stations in the United States in April of 1997. The new assignments were designed to give each television station coverage comparable to the station's existing radio coverage area when they convert to digital transmission.

The change in channel numbers is likely to be a significant challenge for television stations. Many stations, especially in Southern California, have a reduced coverage area. This has resulted in the contesting of the new assignments by some television broadcasters. The Association of Maximum Service Telecasters (an association of local television stations) has proposed an alternative table of channel assignments to address the issues of established broadcasters.

The digital technology that allows high-definition television broadcasts in the U.S. can also be used for "multicasting," that is, transmitting up to five channels of "standard-definition" television programming. Many broadcasters are examining multicasting as an alternative to high-definition television. If the ability to provide more video channels is more desirable than providing high-definition broadcast quality video, HDTV broadcast service and products may be delayed for their entry in the US marketplace.

In July 1996, WRAL in Raleigh, North Carolina became the first United States television station to commence broadcast of high-definition television signals. As of early 1998, more than a dozen stations have licenses for digital transmission, and the number of licenses is increasing every month. HDTV is likely only to be available in the largest markets in the United States for at least the first year the service is provided, so viewers in smaller markets may have to wait many years before they have the opportunity to use digital and/or high-definition television receivers.

The data transmission rate of the HDTV system is 19 Mbps for MPEG-2 video compression and approximately 6-8 Mbps fro MPEG-4 AVC compression format. To allow for a gradual migration to HDTV service, HDTV transmission will also contain regular programming of standard television on HDTV radio channels. The simulcast transmission will continue for up to 15 years as standard NTSC televisions and transmitting facilities are phased out. Initially, HDTV receivers will have the capability to receive and display regular NTSC broadcasts.

The specifications for HDTV digital systems allow for many types of data services in addition to digital video service. Digital HDTV channels carry high-speed digital services that can be addressed to a specific customer or group of customers that are capable of decoding and using those services. Examples of these services include: special programming information, software delivery, video or audio delivery (like pay-per-view programming), and instructional materials.

The data rate available for additional services is dynamic and ranges from a few kbps to several Mbps, depending on the video and audio program content. The gross data rate of the HDTV system is 19 Mbps. The amount of this data rate that is used by the HDTV video signal depends on the compression technology. Video data compression produces a data rate that changes dependent on the original video signal. When the video program contains rapidly changing scenes, most of the 19 Mbps signal is required for transmission. If the video signal is not changing rapidly, much of the 19 Mbps can be used for other types of services.

Transmission of the additional services has a lower priority than transmission of the primary program. If the primary service (HDTV) consumes a large part of the data (such as a rapidly changing video action scene), the customer may have to wait for some time prior to receiving large blocks of data.

Cable Telephony

Cable telephony is the providing of telephone services that use CATV systems to initiate, process, and receive voice communications. Cable telephony systems can either integrate telephony systems with cable modem networks (a teleservice) or the cable modem system can simply act as a transfer method for Internet telephony (bearer service). Because of government regulations (restrictions or high operational level requirements) in many countries, some cable operators are delaying the integration of telephone services with cable networks. In either case, cable telephony systems are data telephony systems that include a voice gateway, gatekeeper, and a media interface.

Voice gateway is a network device that converts communication signals between data networks and telephone networks. A gatekeeper is a server that translates dialed digits into routing points within the cable network or to identify a forwarding number for the public telephone network. A multimedia transfer adapter converts multiple types of input signals into a common communications format.

Figure 1.28 shows a CATV system that offers cable telephony services. This diagram shows that a two-way digital CATV system can be enhanced to offer cable telephony services by adding voice gateways to the cable network's head-end CMTS system and media terminal adapters (MTAs) at the residence or business. The voice gateway connects and converts signals from the public telephone network into data signals that can be transported on the cable modem system. The CMTS system uses a portion of the cable modem signal (data channel) to communicate with the MTA. The MTA converts the telephony data signal to its analog audio component for connection to standard telephones. MTAs are sometimes called integrated access devices (IADs).

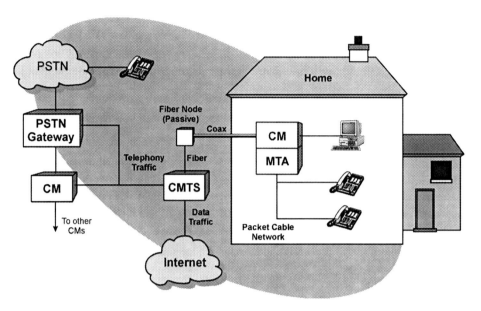

Figure 1.28, Cable Telephony

Because of the high data transmission capability of cable television systems, cable telephony systems can provide video telephony service. Video telephony is a telecommunications service that provides customers with both audio and video signals between their communications devices.

Wireless Cable

"Wireless Cable" is a term given to land based (terrestrial) wireless distribution systems that utilize microwave frequencies to deliver video, data, and/or voice signals to end-users. There are two basic types of wireless cable systems, multichannel multipoint distribution service (MMDS) and local multichannel distribution service (LMDS).

Multichannel video and data services are being offered over microwave frequencies. The data-over-cable service interface specification (DOCSIS) with a few modifications can also be used in 2.6 GHz wireless multipoint, multichannel distribution service (MMDS), and 28 GHz local multipoint distribution service (LMDS) systems [11]. The DOCSIS specification is being adapted for the wireless cable marketplace. A consortium called "Wireless DSL" is working to produce an adapted version of DOCSIS called DOCSIS+ that is suitable for offering cable modem technology via microwave transmission. The DOCSIS+ standard has been proposed to the IEEE 802.16 for conversion into an official standard.

Figure 1.29 shows a LMDS system. This diagram shows that the major component of a wireless cable system is the head-end equipment. The head-end equipment is equivalent to a telephone central office. The head-end building has a satellite connection for cable channels and video players for video on demand. The head-end is linked to base stations (BS) which transmits radio frequency signals for reception. An antenna and receiver in the home converts the microwave radio signals into the standard television channels for use in the home. As in traditional cable systems, a set-top box decodes the signal for input to the television. Low frequency wireless cable systems such as MMDS wireless cable systems (approx 2.5 GHz) can reach up to approximately 70 miles. High frequency LMDS systems (approx 28 GHz) can only reach approximately 5 miles.

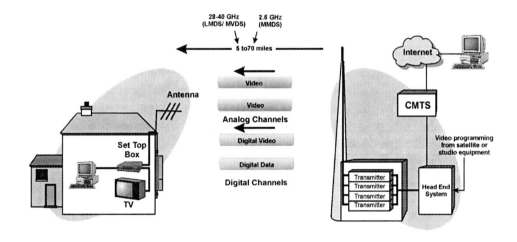

Figure 1.29, Local Multipoint Distribution System (LMDS)

Interactive Television

Interactive television is a combination of cable, television, multimedia, PCs, and network programming that allows dynamic control of media display using inputs from the end-user. Interactive television has three basic types: "pay-per-view" involving programs that are independently billed, "near video-on-demand" (NVOD) with groupings of a single film starting at staggered times, and "video-on-demand" (VOD), enabling request for a particular film to start at the exact time of choice. Interactive television offers interactive advertising, home shopping, home banking, e-mail, Internet access, and games.

Video on demand (VOD) is a service that allows customers to request and receive video services. These video services can be from previously stored media (entertainment movies or education videos) or have a live connection (sporting events in real time).

A limited form of VOD is called near video on demand (NVOD). Near video on demand is a video service that allows a customer to select from a limited number of broadcast video channels. These video channels are typically movie channels that have pre-designated schedule times. Unlike full VOD service, the customer is not able to alter the start or play time of these broadcast videos.

Pay per view (PPV) is a process that allows customers to request the viewing of movies through an unscrambling process. PPV movies are usually broadcasted to all customers in a cable television network. To prevent unauthorized viewing, each PPV channel has its own scrambling code. To provide a customer with a reasonable selection of movies, the same movie is broadcasted on different channels with start intervals that range from 15 to 60 minutes. To provider twenty PPV movies, approximately 80 to 160 television channels would be required.

Analog cable systems provide up to 800 MHz of bandwidth. Using 6 MHz wide video channels, this allows up to 120 analog video channels. By digitizing each 6 MHz channel and using compressed digital video (10:1 compression), this increases the capacity of a cable system to over 500 digital television channels.

Figure 1.30 shows a video on demand system. This diagram shows that multiple video players are available and these video players can be access by the end customer through the set-top box. When the customer browses through the available selection list, they can select the media to play.

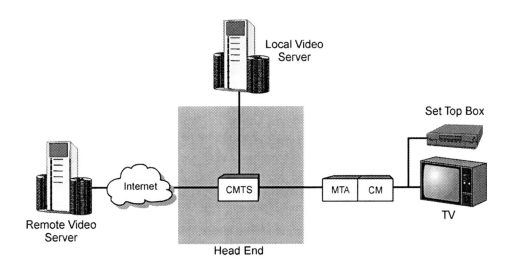

Figure 1.30, Video on Demand (VOD)

Hypervideo is a video program delivery system that allows the embedding of links (hotspots) inside a streaming video signal. This allows the customer (or receiving device) to dynamically alter the presentation of streaming information. Examples of hypervideo could be pre-selection of preferred advertising types or interactive game shows.

Synchronized television (syncTV) is a video program delivery application that simultaneously transmits hypertext markup language (HTML) data that is synchronized with television programming. Synchronized television allows the simultaneous display of a video program along with additional information or graphics that may be provided by advertisers or other information providers.

Internet Protocol Television (IPTV)

Internet protocol television (IPTV) is the process of providing television (video and/or audio) services through the use Internet protocol (IP) networks. These IP networks initiate, process, and receive voice or multimedia communications using IP protocol. These IP systems may be public IP systems (e.g. the Internet), private data systems (e.g. telephone system DSL network), or a hybrid of public and private systems.

Systems

There are several systems that are used for video distribution. The system (standards) used in CATV systems include: NTSC, PAL, SECAM, MPEG and DOCSIS.

National Television Standards Committee (NTSC)

The NTSC system is an analog video system that was developed in the United States and is used in many parts of the world. The NTSC system uses analog modulation where a sync burst precedes the video information. The NTSC system uses 525 lines of resolution (42 are blanking lines) and has a pixel resolution of approximately 148k to 150k pixels.

The NTSC system uses 6 MHz wide radio channels that range from 54 MHz to 88 MHz (for VHF channels 1-6), 174 MHz to 216 MHz (for VHF channels 7-13) and 470 MHz to 806 MHz (for UHF channels 14-69). Initially, the frequency range of 806 MHz to 890 MHz was available for UHF channels 70 to 83. The FCC reallocated these channels for cellular and specialized mobile radio (SMR) use in 1983.

When used in the United States, NTSC systems have a maximum transmitter power level that varies from 100 kWatts for low VHF channels (1-6), 316 kWatts for high VHF channels (7-13) to 5 million Watts for UHF channels (14-69). Television transmission limits are also established based on the class of service (local or wide area) for the authorized television broadcast company.

Figure 1.31 demonstrates the operation of the basic NTSC analog television system. The video source is broken into 30 frames per second and converted into multiple lines per frame. Each video line transmission begins with a burst pulse (called a sync pulse) that is followed by a signal that represents color and intensity. The time relative to the starting sync is the position on the line from left to right. Each line is sent until a frame is complete and the next frame can begin. The television receiver decodes the video signal to position and control the intensity of an electronic beam that scans the phosphorus tube ("picture tube") to recreate the display.

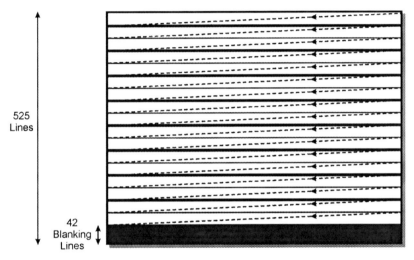

Figure 1.31, NTSC (Analog) Video

Phase Alternating Line (PAL)

The PAL system was developed in the 1980's to provide a common television standard in Europe. PAL is now used in the Middle East and parts of Asia and Africa. The PAL system uses a phase alternation process to enhance the video signal's resistance to chromatic distortions as compared with the NTSC video signal. Although PAL and NTSC systems are similar in function, they are not compatible. A converter box is required between the two systems.

The system provides 625 lines per frame and 50 frames per second. A modified version of PAL (PAL-M) is used for the Brazilian television system. PAL-M provides 525 lines per frame and 60 frames per second. The PAL system uses 7 or 8 MHz wide radio channels.

Figure 1.32 demonstrates the operation of the basic PAL analog television system. The video source is broken into 30 frames per second and converted into 625 lines per frame. Each video line transmission begins with a burst pulse (called a sync pulse) that is followed by a signal that represents color and intensity. The time relative to the starting sync is the position on the line from left to right. Each line is sent until a frame is complete and the next frame can begin. The television receiver decodes the video signal to position and control the intensity of an electronic beam that scans the phosphorus tube ("picture tube") to recreate the display.

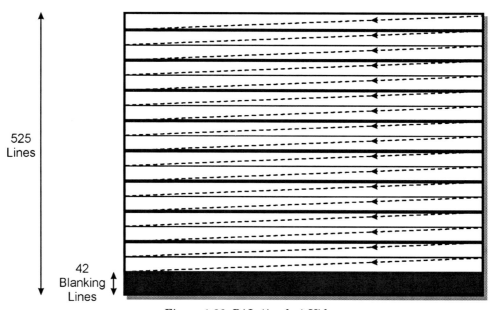

Figure 1.32, PAL (Analog) Video

Sequential Couleur Avec Memoire (SECAM)

Sequential Couleur Avec Memoire (SECAM) is a video transmission system that was developed by France and the former Union of Soviet Socialist Republics to improve on the NTSC video transmission system. This translates to "sequential color with memory." SECAM is a color video transmission system that provides 625 lines per frame and 50 frames per second. This system transfers color difference information sequentially on alternate lines as a FM signal. The SECAM system requires 8 MHz of bandwidth.

Motion Picture Experts Group (MPEG)

Motion picture experts group (MPEG) standards are digital video encoding processes that coordinate the transmission of multiple forms of media (multimedia). Motion picture experts group (MPEG) is a working committee that defines and develops industry standards for digital video systems. These standards specify the data compression and decompression processes and how they are delivered on digital broadcast systems. MPEG is part of International Standards Organization (ISO).

There are various levels of MPEG compression; MPEG-1 and MPEG-2. MPEG-1 compresses by approximately 52 to 1. MPEG-2 compresses up to 200 to 1. MPEG-2 ordinarily provides digital video quality that is similar to VHS tapes with a data rate of approximately 3.2 Mbps. MPEG-2 compression can be used for HDTV channels, however this requires higher data rates.

Figure 1.33 shows how MPEG systems have evolved over time. This diagram shows that the original MPEG specification (MPEG-1) developed in 1991 offered medium quality digital video and audio at up to 1.2 Mbps, primarily sent via CD ROMs. This standard evolved in 1995 to become MPEG-2 that was used for satellite and cable digital television along with DVD distribution. The MPEG specification then evolved into MPEG-4 in 1999 to permit multimedia distribution through the Internet. This example shows that work continues with MPEG-7 for object based multimedia and MPEG-21 for digital rights management.

Introduction to Cable Television (CATV), 2nd Edition

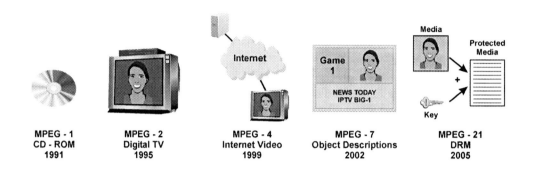

Figure 1.33, Evolution of MPEG

Data Over Cable Service Interface Specifications (DOCSIS)

The data over cable service interface specifications (DOCSIS) is a standard used by cable systems for providing Internet data services to users. The DOCSIS standard was primarily developed by equipment manufacturers and CATV operators. It details most aspects of data over cable networks including physical layer (modulation types and data rates), medium access control (MAC), services, and security. The DOCSIS cable modem specifications are available from CableLabs® at http://www.cablemodem.com/specifications.html.

The downstream information flows to all users that are tuned to a specific RF channel on the cable system. There may be several RF channels used to serve many cable modem users in a system. Each individual cable modem decodes their portion of the data on a specific RF channel. For transmitting on the upstream side, each user is assigned time of a few milliseconds each where the user can transmit short bursts of data. Dividing the channel into small slices of data is well suited for short delays to keyboard commands.

To convert the Internet data into a format suitable for delivery on a cable channel, a CATV upconverter is used at the head-end of the cable system. The CATV upconverter handles both digital and analog television signals.

Usually 10-20 upconverters are installed into a single equipment chassis. To allow cable modems to connect to data networks (such as the Internet), a cable modem termination system (CMTS) is used. The CMTS is an interface device (gateway) that is located at the head-end of a cable television system to send and adapt data between cable modems and other networks.

A single 6 MHz wide television channel is capable of 30-40 Mbps data transmission capacity. This is because coaxial cable offers a communication medium that is relatively noise free (compared to radio or unshielded twist pair cable) that allows the use of complex modulation technologies (combination of amplitude and phase modulation). These modulation technologies can transfer several bits of data for each Hertz of bandwidth (bits per Hertz). In 2001, cable modems could transmit data using 64 QAM modulation technology. To increase the data rate, even more complex modulation technologies such as 256 QAM or even to 1024 QAM have been demonstrated [12].

The DOCSIS system is focused around packet service such as Internet Protocol (IP) and asynchronous transfer mode (ATM) to provide a variety of services (e.g., variable bit-rate, constant bit-rate) with the ability to offer varied levels of quality of service (QoS). This allows the DOCSIS system to offer multiple channels to a home or business that can provide for various services such as voice (constant bit-rate), data (high reliability), and video (high-speed data).

PacketCable™

The name of the project to define a suite of interoperability specifications to allow for devices within a packetized telephony-over-cable network to function correctly even if provided by many vendors.

PacketCable delivers services using standard IP packet data communication over managed IP network. The PacketCable system can sit on top of the DOCSIS system along cable companies to transform their cable data networks into advanced service PacketCable systems.

The PacketCable system is based on IP multimedia system that is an enhancement of session initiation protocol (SIP). The PacketCable system is able to provide voice, data and video services over the IP network while assigning and managing different levels of quality of service (QoS).

The PacketCable system has evolved from a basic packet data communication system to an advanced high-speed multimedia communication system. The initial PacketCable 1.0 core services that allowed for cable telephony (VoIP) services. A PacketCable Multimedia version was released to provide mixed media (multimedia) services. The PacketCable 2.0 version merging of voice, data and video services using modular system architecture. PacketCable 3.0 adds RF channel combining (channel bonding) to increase the maximum data transmission rate to 160 Mbps downstream and 120 Mbps upstream and adds the capability to support IP version 6 addressing (128 bit IP addresses).

OpenCable™

OpenCable is a communication system industry standard that is managed by CableLabs that is designed to allow for advanced television and data communication services. The OpenCable system has a reference design with standard interfaces including headend, core network and consumer product interfaces.

Services

Services that are offered by cable system operators include television distribution, pay per view, advertising, high-speed data services, telephone services.

Television Programming

Television programming involves the sending of television program to groups of consumers that are connected to the cable television network.

Television programs range from free programs (public service and advertiser paid) to premium programming (subscription paid and pay per view) services.

Television service rate plans typically include at least three levels (tiers) of services. These include basic service, mid-level and premium services. The basic service rate plans typically includes several local and regional television programs such as news, weather, community and other relatively low cost programming. The mid-level service plans often add one or two groups of higher value channels such as sports and some movie channels. Premium services typically include several groups of premium channels and 30 to 45 music channels. Some of the key rate plan differences include the individual pricing of channels, enhanced navigation options and an increase in the number of international channels.

Figure 1.34 shows sample cable TV rates throughout Asia, Europe and the USA available in 2006. This table shows that the television service rates are often divided into basic, mid-level and premium groups and that the number of channels offered ranges from approximately 70 channels for basic service to over 160 channels for premium services.

Company	System Type	Basic	Mid-Level	Premium
Charter Communications, Allendale, MI-USA	Cable	$54.99 (76 chan + 1 groups)	$65.99 (76 chan + 2 groups)	$70.99 (76 chan + 1 group + all premium)
Cable TV - Hong Kong Hong Kong[2] (exchange rate used =.13)	Cable	$40.04 (63 chan)	$49.14 (70 chan)	$59.94 (76 chan)
NTL Cable-UK (exchange rate used = 1.89)	Cable	$7.04 (top 10 UK chan+)	$14.72 (100 chan)	$24.96 (160 chan)

Notes: 1. Cable TV Hong Kong rates calculated by adding 1 program group, other program groups were available

Figure 1.34, CATV Service Rate Comparison

Some of the fees associated with cable television service include an installation fee, equipment rental fees, deposits, monthly service fees and pay per view fees. The installation fee typically ranges from $30 to $150. There are various options for waiving (removing) the installation fee ranging from having another service, special promotions or some other event that encourages the customer to take immediate action.

Equipment rental fees for the set top box ranged from $5 to $30 per month. There is typically no rental fee charged for the broadband modem. Equipment deposits range from around $0 to $300. The amount of the deposit is affected by the length of the service contract (monthly, 1 year+). Pay per view fees range from around $1 to $5 per view. Special viewing events could be considerably higher than standard PPV charges.

Figure 1.35 shows a sample cable television service rate plan. This table shows that cable systems typically charge a setup fee of $30 to $150 and have equipment rental fees ranging from $5 to $30 per month. Some cable service providers do not require (waive) deposits and some do require deposits to ensure equipment is returned and not abused. Fees for IPTV monthly services ranged from $13 per month to more than $128 per month. Pay per view fees (PPV) typically cost $1 to $5 per view.

Fee	Cost
Installation	$30-$150
Equipment Rental	$5 to $30
Deposits	$0 to $300
Monthly Service Fees	$13 to $128
Pay per View (typical)	$1 to $5

Figure 1.35, IPTV Service Rate Samples

Pay per View (PPV)

Pay-per-view is a video signal subscription service that allows customers to pay for individual video selections they desire to view. Pay-per-view service can be limited to viewing pre-scheduled broadcasted channels or video on demand (VOD) videos. The typical charge for pay per view services is approximately $6.00.

Advertising

Advertising is the communication of a message or media content to one or more potential customers. Cable operators can make more money from advertising services than they receive from subscription and pay per view services.

The key types of cable television advertising include network advertising, local advertising, addressable advertising and Interactive advertising. Network advertising is advertising messages (adverts) that are provided by a network content provider to local broadcasters. Local broadcasting companies (network affiliates) may receive some payment for the network for the ad insertions or it may receive the television program (network program) at reduced or no cost. Local ad insertion is the selling of ads to advertisers for insertion in that local market (e.g. local businesses).

Addressable advertising is the communication of a message or media content to a specific device or customer based on their address. The address of the customer may be obtained by searching viewer profiles to determine if the advertising message is appropriate for the recipient. The use of addressable advertising allows for rapid and direct measurement of the effectiveness of advertising campaigns. Interactive advertisements are media communication messages that allow a viewer to influence (control) or respond to the advertising message.

Company spending on advertising in all forms of media (television, radio, magazines, and Internet) increases each year with gross national product (GNP). However, the percentage of advertising spent on traditional broadcast television is declining. At the same time the percentage spent on measurable Internet advertising is increasing. According to the Television Bureau of Advertising (www.tvb.org), $10.8 billion was spent on television advertising in the United States during the 3rd quarter of 2006 which was an increase of 3.8% from the same time period in 2005. According to Price Waterhouse Coopers and the Internet Advertising Bureau (www.iab.org), interactive advertising grew by 36% over the same time period, reaching $4.1 billion.

Figure 1.36 shows that television advertising revenue in the United States has increased from approximately $49 billion in 1998 to almost $68 billion in 2005. This chart shows that the portion of advertising spent on television as compared to other media has remained approximately 25% of total advertising expenditures.

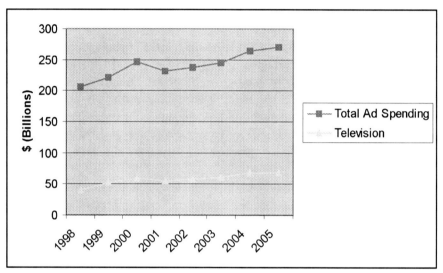

Figure 1.36, Cable Television Advertising in the USA

High Speed Data (Cable Modems)

High-speed data service via cable modems provides customers with the ability to transfer data at broadband data transmission rates (1 Mbps or above). Customers usually pay a monthly fee for high-speed data connection in addition to an Internet service provider (ISP) account.

Cable modem service rates can vary based on the type of use of business or residential, combined service discounts, data transfer rate and data transfer limits. Business service rates can be 2 to 10 times higher than residential service rates. Cable modem service rates are commonly lower for customers who subscribe to other services (bundled) through the cable system (such as television or telephony services). Service plans start with a basic data transfer rate (e.g. 2 Mbps) and a premium fee may be charged to increase the data transfer rate. Most cable modem service plans offer unlimited amounts of data transfer.

The basic data transfer rates are increasing as a result of competition from other broadband service providers. While the data transfer limits of many cable systems are unlimited in 2007, this may change as a result of the availability of television through the Internet service. It is possible (and likely) that some users may transfer 2 to 10 Gigabytes a day (about 2 GB for each our of standard definition quality television through the Internet).

Cable modem services commonly include the assignment of email addresses and an amount web storage area for web site hosting. Business packages commonly include higher data transfer rate options along with an option to assigned a static IP address (important for connecting web servers). Because business services may require that the cable company install and configure systems in areas that do not have established cable facilities (such as downtown commercial business areas), business services may require a 1 to 2 year service contract.

Figure 1.37 shows sample rate plan for high-speed cable modem data access. This chart shows that cable modem access cost varies if the service is for residential or business use. The residential service rates vary if it is included as part of a television service package or if it is used independent of other services. The business packages include basic and premium services where the premium service has higher data transmission rates and more web storage. The residential service plans do not require a service contract while the business plans do require a service contract of 1 to 2 years in this example. This service rate plan does not charge the customer an activation fee if they are adding data service tot heir television rate plan. The business package allows the company to obtain a static (unchanging) IP address.

	Residential		Business	
	With TV Service	Independent	Basic	Premium
Monthly Access	$39.99	$49.99	$89.99	$179.99
Activation	$0	$50	$100	$0
Downlink Speed	2 Mbps	2 Mbps	2 Mbps	4 Mbps
Uplink Speed	256 kbps	256 kbps	256 kbps	512 kbps
Email Addresses	3	3	10	100
Web Storage	10 MB	10 MB	500 MB	2 GB
Static IP Address	N/A	N/A	$10	$10
Modem and Router Lease	$0	$0	$10	$10
Contract Length	None	None	1 Year	2 Years

Figure 1.37, Cable Modem Service Rates

CATV Telephone Services

Cable system telephone services allow customers to make and receive telephone calls through their cable system. Cable telephone services commonly require the installation of a cable telephone adapter box. Some telephone companies do not have the capability of transferring (porting) telephone numbers and this would require a cable telephone customer to obtain a new telephone number.

Cable telephone service rates can vary based on the type of use (business or residential), combined service discounts, calling rate plans and call processing features. Business service rates (if available) can be several times higher than residential service rates. Cable telephone service rates can be lower for customers who subscribe to other services (bundled) through the cable system (such as data modem services).

Cable telephone calling rate plans start with a basic service rate (a telephone line) and additional premium calling rate plans may be added to the basic plan. Some cable telephone service plans offer unlimited domestic long distance calling and discounted international calling rates. Cable telephone service plans commonly include a bundle of call processing services such as call waiting, caller identification and call transfer. Some systems charge a premium for advanced call processing features such as voice mail.

In general, cable telephone service rate costs have been decreasing due to competition from other telephone service providers such as Internet telephone and mobile telephone providers.

Figure 1.38 shows sample rate plan for cable telephone service. This chart shows that cable telephone access cost varies if the service is for residential or business use. The residential service rates vary if it is included as part of a television service package or if it is used independent of other services. The business packages include basic and premium services where the premium service has lower cost of domestic and international calls.

	Residential		Business	
	With TV Service	Independent	Basic	Premium
Monthly Access	$24.99	$34.99	$49.99	$79.99
Activation	$0	$50	$100	$0
Domestic Long Distance	$0	5 cents/min	5 cents/min	$0
International Long Distance	Varies by Country	Varies by Country	Varies by Country	5 cents/min certain countries
Call Features	Included	Included	Included	Included
Voice Mail	$3	$3	$10	$0

Figure 1.38, CATV Telephone Rate Plans

References:

1. "How Cable Television Works," www.Howstuffworks.com, 7 November, 2006.
2. Ibid.
3. Ibid.
4. Ibid.
5. Ibid.
6. Supercomm 2001, "Fujitsu Presentation", 8 June 2001, Atlanta, GA.
7. "How Cable Television Works," www.Howstuffworks.com, 22 November, 2006.
8. "The Worldwide Market for Cable Television Services," www.Instat.com, In-Stat, November 2006
9. www.CATV.org, 12 Sep 01, Statistics, Comsys
10. "How Cable Television Works," www.Howstuffworks.com, 2 August, 2001
11. Wilson, Eric and Shirali, Chet, "Adapting DOCSIS for Broadband Wireless-Access Systems," http://www.csdmag.com

Appendix 1

CATV Acronyms

60i-60 Interlaced
ASN-Abstract Syntax Notation
AWT-Abstract Widowing Toolkit
ADPCM-Adaptive Differential Pulse Code Modulation
AAC-Advanced Audio Codec
AAF-Advanced Authoring Format
AES-Advanced Encryption Standard
ASP-Advanced Simple Profile
ASF-Advanced Streaming Format
ASF-Advanced Systems Format
ATA-Advanced Technology Attachment
ATVEF-Advanced Television Enhancement Forum
ATSC-Advanced Television Systems Committee
Avail-Advertising Availability
Ad Credits-Advertising Credits
Ad Model-Advertising Model
AoD-Advertising on Demand
Ad Splicer-Advertising Splicer
ADET-Aggregate Data Event Table
AEIT-Aggregate Event Information Table
ATIS-Alliance for Telecommunications Industry Solutions
ANSI-American National Standards Institute
ASCII-American Standard Code for Information Interchange
Analog HDTV-Analog High Definition Television
APS-Analog Protection System
ATV-Analog Television
ADC-Analog to Digital Converter
AEE-Application Execution Environment
AIR-Application Information Resource
AIT-Application Information Table
API-Application Program Interface
ARIB-Association for Radio Industries and Business
ADSL-Asymmetric Digital Subscriber Line
AJAX-Asynchronous Javascript and XML
ASI-Asynchronous Serial Interface
ATM-Asynchronous Transfer Mode
AALn-ATM Adaptation Layer n
AC3-Audio Compression 3
AES-Audio Engineering Society
AV-Audio Video
AVI-Audio Video Interleaved
A/V-Audio Visual
AFI-Authority and Format Identifier
AUX-Auxiliary
AVT-AV Transport Service
BIOP-Basic Input Output Protocol
BIFF-Binary Interchange File Format
bslbf-Bit Serial Leftmost Bit First
BMP-BitMaP
BPS-Bits Per Second
BGP-Border Gateway Protocol
BLC-Broadband Loop Carrier
Broadband TV-Broadband Television
BCG-Broadcast Cable Gateway

BFS-Broadcast File System
BML-Broadcast Markup Language
BTSC-Broadcast Television Systems Committee
BTA-Broadcasting Technology Association in Japan
BUC-Buffer Utilization Control
BOF-Business Operations Framework
B2B-Business to Business
B2B Advertising-Business to Business Advertising
B2C-Business to Consumer
CableCARD-Cable Card
CMTS-Cable Modem Termination System
CATV-Cable Television
CTAB-Cable Television Advisory Board
CVCT-Cable Virtual Channel Table
CRLF-Carriage Return Followed by a Line Feed
CSS-Cascading Style Sheets
CUTV-Catch Up Television
CRT-Cathode Ray Tube
CA-Central Arbiter
CPU-Central Processing Unit
CA-Certificate Authority
CRL-Certificate Revocation List
CTL-Certificate Trust List
CDATA-Character Data
CG-Character Generator
CC-Closed Caption
CCTV-Closed-Circuit Television
Coax Amp-Coaxial Amplifier
COFDM-Coded Orthogonal Frequency Division Multiplexing
CLUT-Color Look-Up Table
CVBS-Color Video Blank and Sync
CCIR-Comite' Consultatif International de Radiocommunications
CDR-Common Data Representation
CI-Common Interface
COM-Common Object Model

CORBA-Common Object Request Broker Architecture
CSP-Communication Service Provider
CDA-Compact Disc Audio
CSM-Component Splice Mode
CPL-Composition Playlist
CA-Conditional Access
CAT-Conditional Access Table
CM-Configuration Management
CM-Connection Manager Service
CERN-Conseil European pour la Recherche Nucleaire
CBR Feed-Constant Bit Rate Feed
CE-Consumer Electronics
CEA-Consumer Electronics Association
CGM-Consumer Generated Media
CDS-Content Directory Service
CDA-Content Distribution Agreement
CRID-Content Reference Identifier
CSS-Content Scramble System
CSRC-Contributing SouRCe
CP-Control Point
Co-op-Cooperative Advertising
CN-Core Network
CCBS-Customer Care And Billing System
CSS-Customer Support System
CRC-Cyclic Redundancy Check
DAU-Data Access Unit
DCG-Data Channel Gateway
DDI-Data Driven Interaction
DES-Data Elementary Stream
DEBn-Data Elementary Stream Buffer
DEBSn-Data Elementary Stream Buffer Size
DES-Data Encryption Standard
DET-Data Event Table
DOCSIS®-Data Over Cable Service Interface Specification
DOCSIS+-Data Over Cable Service Interface Specification Plus

DSI-Data Service Initiate
DST-Data Service Table
DDE-Declarative Data Essence
DTS-Decode Time Stamp
DCH-Dedicated Channel
DER-Definite Encoding Rules
DWDM-Dense Wave Division Multiplexing
DDL-Description Definition Language
DCM-Device Control Module
DIB-Device Independent Bitmap
DH-Diffie Hellman
DAVIC-Digital Audio Video Council
DCI-Digital Cinema Initiative
DECT-Digital Enhanced Cordless Telephone
DHN-Digital Home Network
DMA-Digital Media Adapter
DMC-Digital Media Controller
DMD-Digital Media Downloader
DPX-Digital Picture eXchange
DPI-Digital Program Insertion
DRM-Digital Rights Management
DSNG-Digital Satellite News Gathering
DSS-Digital Satellite System
DSP-Digital Signal Processor
DSA-Digital Signature Algorithm
DSS-Digital Signature Standard
DSM-Digital Storage Media
DSM-CC-Digital Storage Media Command and Control
DSM-CC-OC-Digital Storage Media-Command and Control Object Carousel
DSM-CC-UU-Digital Storage Media-Command and Control User to User
DTV-Digital Television
DTT-Digital Terrestrial Television
DTS-Digital Theater Sound
DTS-Digital Theater Systems
DAC-Digital To Analog Converter
DV25-Digital Video 25

DVB-Digital Video Broadcast
DVBSI-Digital Video Broadcast Service Information
DVB-ASI-Digital Video Broadcast-Asynchronous Serial Interface
DVB-MHP-Digital Video Broadcasting Multimedia Home Platform
DV Camcorder-Digital Video Camcorder
DVD-Digital Video Disc
DVE-Digital Video Effect
DVR-Digital Video Recorder
DVS-Digital Video Service
DVI-Digital Visual Interface
Digitizing Tablet-Digitizing Pad
DCC Table-Direct Channel Change Table
DMA Engine-Direct Memory Access Engine
DTC-Direct to Consumer
DTH-Direct To Home
DCC-Directed Channel Change
DCCT-Discrete Channel Change Table
DCT-Discrete Cosine Transform
DFT-Discrete Fourier Tranform
DLP-Discrete Logarithm Problem
DCD-Document Content Description
DOM-Document Object Model
DTD-Document Type Definition
DDB-Download Data Block
DII-Download Info Indication
DCAS-Downloadable Conditional Access System
Drop Amp-Drop Amplifier
DMIF-DSM-CC Multimedia Integration Framework
DASE-DTV Application Software Environment
Dub-Dubbing
DHCP-Dynamic Host Configuration Protocol
DHTML-Dynamic Hypertext Markup Language

DRAM-Dynamic Random Access Memory
EDL-Edit Decision List
EEPROM-Electrically Erasable Programmable Read Only Memory
EFF-Electronic Frontier Foundation
EPF-Electronic Picture Frame
EPG-Electronic Programming Guide
E-Tailers-Electronic Retailers
E-Wallet-Electronic Wallet
ES-Elementary Stream
ESCR-Elementary Stream Clock Reference
EAS-Emergency Alert System
EBS-Emergency Broadcast System
EKE-Encrypted Key Exchange
EDTV-Enhanced Definition Television
EDCA-Enhanced Distributed Channel Access
ECM-Entitlement Control Messages
EFS-Error Free Seconds
EBU-European Broadcasting Union
ECMA-European Commerce Applications Script
ECMA-European Computer Manufactures Association
Euro-DOCSIS-European Data Over Cable Service Interface Specification
EFTA-European Free Trade Association
ETSI-European Telecommunications Standards Institute
EIT-Event Information Table
EOD-Everything on Demand
EE-Execution Engine
EDS-Extended Data Services
ETM-Extended Text Message
ETT-Extended Text Table
EUMID-Extended Unique Material Identifier
FCC-Federal Communications Commission

FIPS-Federal Information Processing Standards
FDDI-Fiber Distributed Data Interface
FTTC-Fiber To The Curb
FTP-File Transfer Protocol
FW-Firmware
FDC-Forward Data Channel
Forward OOB-Forward Out of Band Channel
FourCC-Four Character Code
FOD-Free on Demand
FOSS-Free Open Source Software
FTA-Free to Air
FES-Front End Server
FPA-Front Panel Assembly
FAB-Fulfillment, Assurance, and Billing
GXF-General eXchange Format
GCT-Global Color Table
GPS-Global Positioning System
GSM-Global System For Mobile Communications
GUI-Graphic User Interface
GIF-Graphics Interchange Format
GP-Graphics Processor
HAL-Hardware Abstraction Layer
HCCA-HCF Coordination Channel Access
HITS-Headend in the Sky
HMS-Headend Management System
HSM-Hierarchical Storage Management
HD-High Definition
HD-PLC-High Definition Power Line Communication
HDTV-High Definition Television
HANA-High-Definition Audio-Video Network Alliance
HAVi-Home Audio Video Interoperability
HomePNA-Home Phoneline Networking Alliance
HVN-Home Video Network
HomePlug AV-HomePlug Audio Visual
HMS-Hosted Media Server

HCCA-Hybrid Coordination Function Controlled Channel Access
HFC-Hybrid Fiber Coax
HSTB-Hybrid Set Top Box
HTML-Hypertext Markup Language
HTTP-Hypertext Transfer Protocol
HTTPS-Hypertext Transfer Protocol Secure
IPPV-Impulse Pay Per View
IHDN-In-Home Digital Networks
IR Blaster-Infrared Blaster
IR Receiver-Infrared Receiver
IWS-Initial Working Set
Inline Amp-Inline Amplifier
InstanceID-Instance Identifier
IEEE-Institute Of Electrical And Electronics Engineers
IC-Integrated Circuit
IDE-Integrated Development Environment
IDE-Integrated Drive Electronics
IRD-Integrated Receiver and Decoder
IRT-Integrated Receiver and Transcoder
ISDN-Integrated Services Digital Network
IVG-Integrated Video Gateway
IPR-Intellectual Property Rights
IS-Intensity Stereo
IIC-Inter-Integrated Circuit Bus
ICG-Interactive Cable Gateway
ICAP-Interactive Communicating Application Protocol
IPG-Interactive Programming Guide
IDL-Interface Definition Language
IF-Intermediate Frequency
IF Switching-Intermediate Frequency Switching
ICC-International Color Consortium
IDEA-International Data Encryptions Algorithm
ISAN-International Standard Audiovisual Number
ISBN-International Standard Book Number

ISO-International Standards Organization
ITU-International Telecommunication Union
Net-Internet
IANA-Internet Assigned Numbering Authority
IETF-Internet Engineering Task Force
IGMP-Internet Group Management Protocol
IIOP-Internet Inter-ORB Protocol
ISMA-Internet Media Streaming Alliance
IP-Internet Protocol
IPCATV-Internet Protocol Cable Television
IPDC-Internet Protocol Datacasting
IP STB-Internet Protocol Set Top Box
IPTV-Internet Protocol Television
ISP-Internet Service Provider
iTV-Internet TV
IOR-Interoperable Object Reference
IDCT-Inverse Discrete Cosine Transform
IPVBI-IP Multicast over VBI
IMS-IP Multimedia System
IIF-IPTV Interoperability Forum
JAR-Java Archive
JAAS-Java Authencation and Authroization Service
JCA-Java Cryptography Architecture
JCE-Java Cryptography Extentions
JDK-Java Development Kit
JMF-Java Media Framework
JNI-Java Native Interface
JNM-Java Native Methods
JSSE-Java Secure Socket Extension
JVM-Java Virtual Machine
JCIC-Joint Committee on Intersociety Coordination
JPEG-Joint Photographic Experts Group
JS-Joint Stereo
JFIF-JPEG File Interchange Format
JNG-JPEG Network Graphics
KoD-Karaoke on Demand

KDF-Key Derivation Function
KLV-Key Length Value
KPI-Key Performance Indicator
KQI-Key Quality Indicators
Killer App-Killer Application
kbps-Kilo bits per second
LBI-Late Binding Interface
Linear TV-Linear Television
LCD-Liquid Crystal Display
Liquid LSP-Liquid Label Switched Path
LISP-LIS Processing
LAN-Local Area Network
LID-Local Identifier
LMDS-Local Multichannel Distribution Service
LMDS-Local Multipoint Distribution System
LOC-Local Operations Center
LP-2-Local Primary Monitoring Station Alternate
LP-1-Local Primary Monitoring Station First
LCN-Logical Channel Number
LLC-SNAP-Logical Link Control-Sub Network Access Protocol
LSD-Logical Screen Descriptor
MGG-LC-Low Complexity MNG
LFE-Low Frequency Enhancement
LSF-Low Sampling Frequency
MRD-Marketing Requirements Document
MCR-Master Control Room
MGT-Master Guide Table
MXF-Material eXchange Format
MTU-Maximum Transmission Unit
MSE-Mean Square Error
MAC-Media Access Control
MAP-Media Access Plan
MB-Media Block
MDI-Media Delivery Index
MIU-Media Interoperability Unit
MS-Media Server
MAC-Medium Access Control

MMU-Memory Management Unit
MAC-Message Authentcation Code
MMS-Microsoft Media Server Protocol
MRLE-Microsoft Run Length Encoding
Mbps-Millions of bits per second
MG-Minimum Guarantee
MITRE-Missile Test and Readiness Equipment
MTT-Mobile Terrestrial Television
MDCT-Modified Discrete Cosine Transformation
MER-Modulation Error Ratio
MSB-Most Significant Bit
MJPEG-Motion JPEG
MPEG-Motion Picture Experts Group
MP3-Motion Picture Experts Group Layer 3
MP3-Motion Picture Experts Group Level 3
Studio-Movie Studio
MPEGoIP-MPEG over Internet Protocol
MRD-MPEG-2 Registration Descriptor
MP4-MPEG-4
MBONE-Multicast Backbone
MBGP-Multicast Border Gateway Protocol
MSDP-Multicast Source Discovery Protocol
MMDS-Multichannel Multipoint Distribution Service
MVDDS-Multichannel Video Distribution and Data Service
MVPD-Multichannel Video Program Distributor
MCNS-Multimedia Cable Network System
MHP-Multimedia Home Platform
MoCA-Multimedia over Coax Alliance
MHEG-Multimedia/Hypermedia Expert Group
MNG-Multiple image Network Graphics
MCard-Multiple Stream CableCARD
MSO-Multiple System Operator

MuX-Multiplexer
MMDS-Multipoint Microwave Distribution System
MPTS-Multiprogram Transport Stream
MPTS Feed-Multiprogram Transport Stream Feed
MIME-Multipurpose Internet Mail Extensions
MIDI-Musical Instrument Digital Interface
NISDN-Narrrow-band Integration Services Digital Network
NAB-National Association Of Broadcasters
NCTA-National Cable Television Association
NIST-National Institute Of Standards And Technology
NOC-National Operations Center
NTSC-National Television System Committee
NICAM-Near Instantaneous Companded Audio Multiplexing
NVOD-Near Video On Demand
NUT-Net UDP Throughput
NCF-Network Connectivity Function
NI-Network Interface
NIC-Network Interface Card
NOC-Network Operations Center
NPVR-Network Personal Video Recorder
NRT-Network Resources Table
NSAP-Network Service Access Point
NTP-Network Time Protocol
NNW-No New Wires
NPT-Normal Play Time
NAN-Not A Number
OC-Object Carousel
ORB-Object Request Broker
OSD-On Screen Display
On Airwaves-On-Air
OCAP-Open Cable Application Platform
OSPF-Open Shortest Path First
OSGI-Open Systems Gateway Initiative

OS-Operating System
OSS-Operations Support System
OUI-Organization Unique Identifier
OOB Channel-Out of Band Channel
OOB Receiver-Out of Band Receiver
OBE-Out of Box Experience
PDV-Packet Delay Variation
PER-Packet Error Rate
PID-Packet Identifier
PID Dropping-Packet Identifier Dropping
PES-Packetized Elementary Stream
Parametric QoS-Parametric Quality of Serivce
PCDATA-Parsed Character Data
PLTV-Pause Live Television
PPD-Pay Per Day
PPV-Pay Per View
PSNR-Peak Signal to Noise Ratio
PCI-Peripheral Component Interconnect
PCMCIA-Personal Computer Memory Card International Association
PDA-Personal Digital Assistant
PMS-Personal Media Server
PVR-Personal Video Recorder
PAL-Phase Alternating Line
PIG-Picture in Graphics
PIP-Picture in Picture
PSU-Pillow Speaker Unit
PKIX-PKI X.509
POTS-Plain Old Telephone Service
Play-List-Playlist
POD-Point of Deployment
POD Module-Point of Deploymnet
POE-Point of Entry
PFR-Portable Font Resource
PMI-Portable Media Interface
PMP-Portable Media Player
PNG-Portable Network Graphics
PLC-Power Level Control
Preamp-Pre-Amplifier
PE-Presentation Engine
PTS-Presentation Time Stamp

PU-Presentation Unit
PGP-Pretty Good Privacy
Priority Based QoS-Priority Based Quality of Serivce
PEM-Privacy Enhanced Mail
Private TV-Private Television
PCR-Production Control Room
PSIP-Program and System Information Protocol
PAT-Program Association Table
PCR-Program Clock Reference
PMT-Program Map Table
PSI-Program Specific Information
PSM-Program Splice Mode
PSTD-Program System Target Decoder
PMS-Property Management System
PIMDM-Protocol Independent Multicase Dense Mode
PIM-Protocol Independent Multicast
PBS-Public Broadcast Service
PKI-Public Key Infrastructure
PLMN-Public Land Mobile Network
PSTN-Public Switched Telephone Network
PAM-Pulse Amplitude Modulation
PCM-Pulse Coded Modulation
PDM-Pulse Duration Modulation
Push VOD-Push Video on Demand
QAM-Quadrature Amplitude Modulation
QPSK-Quadrature Phase Shift Keying
QoS-Quality Of Service
QoS Policy-Quality of Service Policy
QT Atoms-Quicktime Atoms
MOV-QuickTime MOVie format
RF Bypass-Radio Frequency Bypass Switch
RF Modulator-Radio Frequency Modulator
RF Out-Radio Frequency Output
RAM-Random Access Memory
RRT-Rating Region Table
ROM-Read Only Memory

RTOS-Real Time Operating System
RTP-Real Time Protocol
RTSPT-Real Time Streaming Protocol over TCP
RTSPU-Real Time Streaming Protocol over UDP
RTCP-Real-Time Transport Control Protocol
RSS-Really Simple Syndication
RM-RealMedia
RPTV-Rear Projection Television
RGBA-Red Green Blue Alpha
RGB-Red, Green, Blue,
RPC-Regional Protection Control
RA-Registration Authority
RADIUS-Remote Access Dial In User Service
RAS-Remote Access Server
RC-Remote Control
RDP-Remote Display Protocol
RHVO-Remote Video Hub Operation
RCS-Rendering Control Service
RP-Rendezvous Point
RFC-Request For Comments
RG-Residential Gateway
RDF-Resource Description Framework
RIFF-Resource Interchange File Format
RSVP-Resource Reservation Protocol
RCC-Reverse Control Channel
RDC-Reverse Data Channel
Reverse OOB-Reverse Out of Band Channel
RPF-Reverse Path Forwarding
RAS-Rights Access System
RSA-Rivest, Shamir, Adleman
SBNS-Satellite and Broadcast Network System
SMATV-Satellite Master Antenna Television
SMS-Screen Management System
SAP-Secondary Audio Program
SRTP-Secure Real Time Protocol

SDD-Self Describing Device
S-Video-Separate Video
SECAM-Sequential Couleur Avec MeMoire
SDI-Serial Digital Interface
SCP-Service Control Protocol
SDF-Service Description Framework
SDT-Service Description Table
SDS-Service Discovery and Selection
SGW-Service Gateway
SID-Service Identifier
SI-Service Information
SRM-Session and Resource Manager
SAP-Session Announcement Protocol
SDP-Session Description Protocol
SBB-Set Back Box
SNMP-Simple Network Management Protocol
SCard-Single Stream CableCARD
SCSI-Small Computer Systems Interface
SCTE-Society of Cable Telecommunication Engineers
Softkeys-Soft Keys
SDK-Software Development Kit
S/PDIF-Sony Philips Digital InterFace
SD-Standard Definition
SDTV-Standard Definition Television
SGML-Standard Generalized Markup Language
SII-Station Identification Information
Stereo-Stereophonic
SAN-Storage Area Network
Stylesheet-Style Sheet
SAC-Subscriber Acquisition Cost
SAS-Subscriber Authorization System
SMS-Subscriber Management System
STV-Subscription Television
SVOD-Subscription Video on Demand
SHE-Super Headend
SDV-Switched Digital Video
Sync Impairment-Synchronization Impairments

SPG-Synchronization Pulse Generator
SDH-Synchronous Digital Hierarchy
SRTS-Synchronous Residual Time Stamp
SCR-System Clock Reference
SI-System Information
SMBUS-System Management Bus
SOC-System On Chip
STC-System Time Clock
TV Centric-Television Centric
TV Channel-Television Channel
T-Mail-Television Mail
TVoF-Television over Fiber
TV Portal-Television Portal
TV Studio-Television Studio
TMS-Theater Management System
TSTV-Time Shift Television
TSTV-Time Shifted Television
Timestamp-Time Stamp
TRT-Total Running Time
TFS-Transient File System
TXOP-Transmission Opportunity
TFS-Transport File System
Transport ID-Transport Identifier
TSFS-Transport Stream File System
3DES-Triple Data Encryption Standard
TLV-Type Length Value
UMB-Ultra Mobile Broadband
UMID-Unique Material Identifier
UPnP-Universal Plug and Play
UPnP AV-Universal Plug and Play Audio Visual
Upfronts-Upfront Advertising
UGC-User Generated Content
UI-User Interface
USC-User Selectable Content
USM-User Services Management
V-Factor-V Factor
VBR Feed-Variable Bit Rate Feed
VLE-Variable Length Encoding
VBI-Video Blanking Interval
vBook-Video Book
VBV-Video Buffer Verifier

VDN-Video Distribution Network
VHO-Video Hub Office
VOD-Video On Demand
VQI-Video Quality Index
VQM-Video Quality Measurement
Video Ringtone-Video Ring Tone
VC-Virtual Channel
VCT-Virtual Channel Table
VM-Virtual Machine
Virtual TV Channel-Virtual Television Channel
WTCG-Watch This Channel Grow
WM-Windows Media
WMA-Windows Media Audio
XSLT-XML Stylesheet Language Transformation

Index

Ad Server, 30
Addressable Advertising, 30, 75
Advanced Audio Codec (AAC), 43
Advanced Encryption Standard (AES), 42
Advertiser, 73
Advertising Availability (Avail), 30
Advertising Splicer (Ad Splicer), 29, 31
Analog Cue Tone, 30
Analog Television (ATV), 3-4, 20, 28, 38, 52, 54-55, 67-68, 70
Application Information Resource (AIR), 14, 17, 20
Asset Storage, 24, 28
Asynchronous Transfer Mode (ATM), 19, 71
Audio Channel, 22
Audio Engineering Society (AES), 42
Audio Processing, 42-43
Audio Visual (A/V), 45-46
Automation, 27, 32
Baseband, 47, 57
Billing System, 26
Bits Per Second (BPS), 3
Bookings, 24
Cable Card (CableCARD), 45
Cable Modem Termination System (CMTS), 8, 58, 61, 71
Cable Ready Television, 45
Cable Telephone, 48, 51, 78-79
Cable Telephone Adapters, 48
Cable Television (CATV), 1-80
Caching, 28
Channel, 2-5, 15-25, 29, 34-35, 37, 40, 47, 53, 55-59, 61, 64, 70-72
Channel Combiner, 18, 23
Channel Processor, 22
Chrominance, 52
Closed Caption (CC), 4
Combining Network, 23
Common Object Model (COM), 70, 80
Compliance, 11
Configuration Management (CM), 8
Connection Manager Service (CM), 8
Constant Bit Rate Feed (CBR Feed), 21
Consumer Electronics (CE), 59
Content, 6, 9-14, 16, 20, 24-26, 29, 32, 41, 46, 56, 60, 75
Content Acquisition, 24-26
Content Distribution, 25
Content Encoding, 29
Content Feed, 13
Content Processing, 24, 29
Content Provider, 13, 75
Content Rights Management, 56
Content Source, 14
Contribution Network, 9, 11-12, 16-17
Core Network (CN), 32, 34, 72
Cue Tones, 30
Data Elementary Stream (DES), 42
Data Encryption Standard (DES), 42
Data Over Cable Service Interface Specification (DOCSIS)®, 35, 62-63, 66, 70-71, 80

Data Over Cable Service Interface Specification Plus (DOCSIS+), 35, 62-63, 66, 70-71, 80
Digital Program Insertion (DPI), 30
Digital Rights Management (DRM), 42, 69
Digital Signal Processor (DSP), 42, 45
Digital Storage Media (DSM), 28
Digital Television (DTV), 2, 19-21, 38, 53-56, 58-59, 64, 69
Digital Terrestrial Television (DTT), 20, 46
Digital Video Disc (DVD), 55, 69
Digital Video Service (DVS), 60
Display Device, 45
Distribution Control, 24, 31
Educational Access Channel, 15
Effects, 2, 27
Electronic Programming Guide (EPG), 26
Embedded, 11, 30, 42-43
Embedded Application, 43
Emergency Alert System (EAS), 16
Federal Communications Commission (FCC), 3, 59, 66
Feed, 11, 13-17, 21
Fiber Ring, 32
Fiber Spur, 32
Government Access Channel, 15-16
Graphics, 29, 41-43, 65
Graphics Processing, 29, 42-43
Graphics Processor (GP), 29
Helicopter Feed, 11, 14
High Definition (HD), 19, 41, 52, 55, 59
High Definition Television (HDTV), 4, 52, 55, 59-61, 69

Home Banking, 64
Home Shopping, 64
Hybrid Fiber Coax (HFC), 32
Hybrid Set Top Box (HSTB), 46
Hypertext Markup Language (HTML), 65, 70
Hypertext Transfer Protocol (HTTP), 70, 80
Ingesting Content, 25
Installation, 74, 78
Institute Of Electrical And Electronics Engineers (IEEE), 63
Integrated Circuit (IC), 42
Integrated Receiver and Decoder (IRD), 18-19
Intensity Stereo (IS), 1-5, 7-72, 74-75, 77, 79
Intermediate Frequency (IF), 33-34, 41, 54, 59-61, 71, 75, 77-79
International Standards Organization (ISO), 69
Internet (Net), 1-2, 8, 10-12, 16, 26, 31, 35, 48, 57-58, 61, 64, 66, 69-71, 75-77, 79
Internet Protocol (IP), 11, 19, 29, 48, 66, 71-72, 77
Internet Protocol Television (IPTV), 40, 46, 66, 74
Internet Service Provider (ISP), 76
IP Multimedia System (IMS), 72
Kilo bits per second (kbps), 60
Linear Television (Linear TV), 26-27
Live Feed, 15
Loader, 43
Local Ad Insertion, 75
Local Area Network (LAN), 36-37
Local Feed, 14
Local Headend, 18

Index

Local Multichannel Distribution Service (LMDS), 62-63
Local Multipoint Distribution System (LMDS), 62-63
Media Access Control (MAC), 57, 70
Media Player, 44
Media Processor, 45
Medium Access Control (MAC), 57, 70
Memory, 28, 45-46, 69
Message Authentcation Code (MAC), 57, 70
Metadata Management, 24, 26
Metadata Normalization, 26
Microwave Link, 5, 10
Middleware, 41, 44
Middleware Client, 44
Middleware Compatibility, 41, 44
Millions of bits per second (Mbps), 21, 35, 37-38, 47, 55, 57-58, 60, 69, 71-72, 76
Motion Picture Experts Group (MPEG), 4, 19, 21-22, 31, 34, 41-42, 45, 53, 66, 69-70
Motion Picture Experts Group Layer 3 (MP3), 43
Motion Picture Experts Group Level 3 (MP3), 43
Movie Studio (Studio), 4-6, 14
MPEG-4 (MP4), 29, 60, 69
Multichannel Multipoint Distribution Service (MMDS), 62-63
Multichannel Video Program Distributor (MVPD), 9
Multiple System Operator (MSO), 9
Multipoint Microwave Distribution System (MMDS), 62-63
Multiprogram Transport Stream (MPTS), 19
Must Carry Regulations, 20
National Television System Committee (NTSC), 4, 20, 22, 34, 42, 45, 52, 54, 59-60, 66-67, 69
Near Video On Demand (NVOD), 64
Nearline Storage, 28
Network Access, 34
Network Advertising, 75
Network Content, 75
Network Feeds, 4, 6, 13
Network Interface (NI), 39-40
Off Air Feed, 14
Offline Storage, 28-29
On Screen Display (OSD), 43
Online Storage, 28
OpenCable™, 72
Operating System (OS), 46
Packet Error Rate (PER), 3, 23, 35, 49, 51-53, 64, 67-69, 71-75
Packet Identifier (PID), 22
PacketCable™, 71-72
Packetized Elementary Stream (PES), 22
Pay Per View (PPV), 3, 64, 72-75
Personal Video Recorder (PVR), 44
Phase Alternating Line (PAL), 4, 20, 22, 34, 42, 45, 52, 54, 59, 66-68
Playout, 6, 24-27
Playout System, 26-27
Plug-In, 44
Porting, 78
Primary Event, 27
Production, 4, 31
Program Guide, 40
Programming Sources, 17
Public Access Channel, 15

Public Key Infrastructure (PKI), 42
Quadrature Amplitude Modulation (QAM), 35, 47, 71
Quadrature Phase Shift Keying (QPSK), 35, 47
Quality Of Service (QoS), 36-37, 71-72
Radio Frequency Modulator (RF Modulator), 7
Read Only Memory (ROM), 46
Really Simple Syndication (RSS), 16
Remote Control (RC), 46
Rights Management, 42, 56, 69
Sample Rate, 77, 79
Secure Microprocessor, 42
Sequential Couleur Avec MeMoire (SECAM), 20, 22, 52, 66, 69
Software Module, 44
Special Effects, 27
Squeeze Back, 27
Standard Definition (SD), 41, 55, 77
Standard Definition Television (SDTV), 55
Stereophonic (Stereo), 4, 52
Stored Media, 6, 9, 11-13, 44, 64
Streaming, 26-27, 31, 65
Streaming Video, 65
Studio, 4-6, 14
Super Headend (SHE), 18
Surround Sound, 43
Syndication Feeds, 16
System Information (SI), 35
Television Advertising, 75-76
Television Channel (TV Channel), 2-5, 20, 22, 34, 55, 58-59, 71
Television Set, 41, 43, 52
Television Studio (TV Studio), 5
Transcoder, 20
Transcoding, 29
Truck Feed, 14
Tuner, 20, 35, 41, 45, 47, 57
Tuning Head, 20
Upconverter, 70
Upgradability, 44
Usage Rights, 25
User Interface (UI), 40, 44-46
Variable Bit Rate Feed (VBR Feed), 21
Video Broadcasting, 1
Video Encoder, 20
Video On Demand (VOD), 63-65, 74
Video Processing, 42
Video Warping, 42
Warping, 42
Web Browser, 44
Workflow, 24-25, 32
Workflow Automation, 32
Workflow Management, 24-25

Althos Publishing Book List

Product ID	Title	# Pages	ISBN	Price	Copyright
Billing					
BK7727874	Introduction to Telecom Billing	48	0974278742	$11.99	2004
BK7769438	Introduction to Wireless Billing	44	097469438X	$14.99	2004
Business					
BK7781359	How to Get Private Business Loans	56	1932813594	$14.99	2005
BK7781368	Career Coach	92	1932813683	$14.99	2006
Datacom					
BK7727873	Introduction to Data Networks	48	0974278734	$11.99	2003
IP Telephony					
BK7727877	Introduction to IP Telephony	80	0974278777	$12.99	2003
BK7781361	Tehrani's IP Telephony Dictionary, 2nd Edition	628	1932813616	$39.99	2005
BK7780530	Internet Telephone Basics	224	0972805303	$29.99	2003
BK7780532	Voice over Data Networks for Managers	348	097280532X	$49.99	2003
BK7780538	Introduction to SIP IP Telephony Systems	144	0972805389	$14.99	2003
BK7781311	Creating RFPs for IP Telephony Communication Systems	86	193281311X	$19.99	2004
BK7781309	IP Telephony Basics	324	1932813098	$34.99	2004
BK7769430	Introduction to SS7 and IP	56	0974694304	$12.99	2004
IP Television					
BK7781362	Creating RFPs for IP Television Systems	86	1932813624	$19.99	2005
BK7781357	IP Television Directory	154	1932813578	$89.99	2005
BK7781355	Introduction to Data Multicasting	68	1932813551	$14.99	2005
BK7781340	Introduction to Digital Rights Management (DRM)	84	1932813403	$14.99	2005
BK7781351	Introduction to IP Audio	64	1932813519	$14.99	2005
BK7781335	Introduction to IP Television	104	1932813357	$14.99	2005
BK7781330	Introduction to IP Video Servers	68	1932813306	$14.99	2005
BK7781341	Introduction to IP Video	88	1932813411	$14.99	2005
BK7781352	Introduction to Mobile Video	68	1932813527	$14.99	2005
BK7781353	Introduction to MPEG	72	1932813535	$14.99	2005
BK7781342	Introduction to Premises Distribution Networks (PDN)	68	193281342X	$14.99	2005
BK7781354	Introduction to Telephone Company Television (Telco TV)	84	1932813543	$14.99	2005
BK7781344	Introduction to Video on Demand (VOD)	68	1932813446	$14.99	2005
BK7781356	IP Television Basics	308	193281356X	$34.99	2005
BK7781334	IP TV Dictionary	652	1932813349	$39.99	2005
BK7781363	IP Video Basics	280	1932813632	$34.99	2005
Programming					
BK7727875	Wireless Markup Language (WML)	287	0974278750	$34.99	2003
BK7781300	Introduction to xHTML:	58	1932813004	$14.99	2004
Legal and Regulatory					
BK7769433	Practical Patent Strategies Used by Successful Companies	48	0974694339	$14.99	2003
BK7781332	Strategic Patent Planning for Software Companies	58	1932813322	$14.99	2004
BK7780533	Patent or Perish	220	0972805338	$39.95	2003
Telecom					
BK7727872	Introduction to Private Telephone Systems 2nd Edition	86	0974278726	$14.99	2005
BK7727876	Introduction to Public Switched Telephone 2nd Edition	54	0974278769	$14.99	2005
BK7780537	SS7 Basics, 3rd Edition	276	0972805370	$34.99	2003
BK7780535	Telecom Basics, 3rd Edition	354	0972805354	$29.99	2003
BK7727870	Introduction to Transmission Systems	52	097427870X	$14.99	2004
BK7781313	ATM Basics	156	1932813136	$29.99	2004
BK7781302	Introduction to SS7	138	1932813020	$19.99	2004
BK7781345	Introduction to Digital Subscriber Line (DSL)	72	1932813454	$14.99	2005

For a complete list please visit
www.AlthosBooks.com

Althos Publishing Book List

Wireless

Product ID	Title	Pages	ISBN	Price	Year
BK7781306	Introduction to GPRS and EDGE	98	1932813063	$14.99	2004
BK7781304	Introduction to GSM	110	1932813047	$14.99	2004
BK7727878	Introduction to Satellite Systems	72	0974278785	$14.99	2005
BK7727879	Introduction to Wireless Systems	536	0974278793	$11.99	2003
BK7769432	Introduction to Mobile Telephone Systems	48	0974694320	$10.99	2003
BK7769435	Introduction to Bluetooth	60	0974694355	$14.99	2004
BK7769436	Introduction to Private Land Mobile Radio	50	0974694363	$14.99	2004
BK7769434	Introduction to 802.11 Wireless LAN (WLAN)	62	0974694347	$14.99	2004
BK7769437	Introduction to Paging Systems	42	0974694371	$14.99	2004
BK7781308	Introduction to EVDO	84	193281308X	$14.99	2004
BK7781305	Introduction to Code Division Multiple Access (CDMA)	100	1932813055	$14.99	2004
BK7781303	Wireless Technology Basics	50	1932813039	$12.99	2004
BK7781312	Introduction to WCDMA	112	1932813128	$14.99	2004
BK7780534	Wireless Systems	536	0972805346	$34.99	2004
BK7769431	Wireless Dictionary	670	0974694312	$39.99	2005
BK7769439	Introduction to Mobile Data	62	0974694398	$14.99	2005

Optical

Product ID	Title	Pages	ISBN	Price	Year
BK7781329	Introduction to Optical Communication	132	1932813292	$14.99	2006

Order Form

Phone: 1 919-557-2260
Fax: 1 919-557-2261
404 Wake Chapel Rd., Fuquay-Varina, NC 27526 USA

Date:_____

Name:_____ Title:_____
Company:_____
Shipping Address:_____
City:_____ State:_____ Postal/ Zip:_____
Billing Address:_____
City:_____ State:_____ Postal/ Zip _____
Telephone:_____ Fax:_____
Email: _____
Payment (select): VISA ___ AMEX ___ MC ___ Check ___
Credit Card #: _____ Expiration Date: _____
Exact Name on Card: _____

Qty.	Product ID	ISBN	Title	Price Ea	Total
Book Total:					
Sales Tax (North Carolina Residents please add 7% sales tax)					
Shipping: $5 per book in the USA, $10 per book outside USA (most countries). Lower shipping and rates may be available online.					
Total order:					

For a complete list please visit
www.AlthosBooks.com